TIME
TAMED

To Alexandra, Max and Freddie,
who mitigate the passage of time.

TIME
TAMED

The remarkable story of humanity's
quest to measure time

Nicholas Foulkes

SIMON &
SCHUSTER

London · New York · Sydney · Toronto · New Delhi

A CBS COMPANY

MMXIX

Contents

Introduction

This book is not intended to be a serious technical treatise on clocks and watches, although there are some serious, highly technical clocks and watches within its pages. Neither is it a scholarly, philosophical disquisition on the nature of time (I'll leave that to the really clever people); nor, finally, should it be taken in any way as a definitive account of the millennia man has spent trying to tame time – a task as Sisyphean as it is Canute-like.

If the above paragraph has not dissuaded you from continuing to read, I suggest that you treat this book as you would a collection of very loosely linked short stories. Each chapter can be taken, and I hope enjoyed, on its own, as a discrete narrative; but also read in sequence, strung together like lambent pearls or dazzling precious stones, each one enhancing its neighbours to create a glittering jewel that is more than the sum of its parts. Or, at least, that is the idea – I only wish my writing were worthy of the extravagant simile.

What is in no doubt is that the following selection of twenty-eight timepieces have all, in their way, contributed to history, and will take us on a journey from Mesolithic Scotland to Belle Époque Paris; from the bottom of the Mediterranean to the surface of the moon; from the court of Charlemagne to the cockpit of a Pan Am Boeing; from Jacobean London to eleventh-century China.

But, however disparate the stories may seem, they are linked by the thread of human ingenuity and, more often than not, beauty. A timepiece is often an object in which science and art meet, where the calculation of gearing ratios and the finer aspects of the goldsmith's talents assume equal importance. In my opinion, the perfect timepiece is one that intrigues the mind and delights the eye, and in this book you will meet examples that achieve one, the other, and often both.

An understanding of time is arguably what makes us human; it is unsurprising that we value it and the machines that calibrate it for us. Indeed, the very story of human civilisation can be told through our ever-developing concept of time and the instruments with which we have interpreted it.

In its infancy, mankind experienced time as literally heaven-sent. The Earth's rotation on its axis provided the span of the day, and the period of the Earth's elliptical 365.25-day (give or take) orbit of the sun provided what we call a year. Meanwhile, the moon supplied the observable phenomenon of its waxing and waning, a cycle that took approximately 29.5 days, on which we based the concept of a month. The problem, of course, was that the solar and lunar cycles did not quite coordinate. It is possible that primitive men were grappling with this inconvenience as early as 12,000 years ago.

The Roman emperor Trajan depicted as pharaoh, offering a water clock to the goddess Hathor. Relief of the Mammisi, Temple of Hathor, 88–51 BCE, Dendera, Egypt.

The reconciling of the solar and lunar calendars would occupy mankind for millennia, and continues to do so today. Babylonian astronomers found that these two calendars coincided once every nineteen years, and it was the Greek astronomer Meton of Athens who lent his name to the calendar that was used by Ancient Greece until 46 BCE.

However, while neat, a nineteen-year cycle is long, and all the time men were becoming more accurate in their appreciation of time. In Ancient Babylon, the span of daylight was divided into twelve hours and depicted on sundials, while the Egyptians had learned to tell the time at night using the position of the stars. Our seconds and minutes – the former the sixtieth part of the latter, itself a sixtieth part of a larger unit of time – were derived from sexagesimal counting that became the dominant form of calculation in Mesopotamia around 5000 years ago.

Of course, minutes and seconds were an abstract concept, more mathematical than temporal, and derived from this system came the 360 degrees of longitude that divide the world. (As we will discover, longitude would reunite with the human quest for

Plaster cast of the oldest extant Clepsydra (now held in the Cairo Museum) – the world's first (surviving) accurate timepiece – discovered at the temple of Karnak, dating from 1415 to 1380 BCE.

accuracy in the eighteenth century.) For centuries, mathematics, astronomy, astrology, theology and time remained tied in the most Gordian of philosophical knots.

With the invention of the water clock in Pharaonic Egypt, a way of measuring time (by the amount of liquid that flows regularly from a vessel) had been found. Finally, man had wrested time from the celestial bodies, and slowly the practical application of this almost Promethean gift began. Of course, for the majority, accurate time was not a necessity; ask a medieval peasant the time and he would probably tell you what season it was. Nevertheless, by the early Middle Ages, a combination of water clocks and solar observations were governing activities as varied as the closure of city gates and prayer – all signalled aurally by hand-rung bells.

No one knows who invented the mechanical clock, but its arrival in thirteenth-century Europe brought with it the Renaissance and, a little later, the period of European world domination known as the Age of Discovery. For Karl Marx, its importance was in no doubt: 'The clock is the first automatic machine applied to practical purposes,' he wrote to Engels in 1863, 'and the whole theory of production of regular motion was developed on it.' Even though abstract, time had become the ultimate economic object. For Marx, the factory clock had become a symbol of depersonalization and the commoditization of human effort.

The pursuit of time – and efforts to capture it with machines – has involved strong personalities and great characters who, through their stubborn perseverance, innate genius or sheer eccentricity, have written their names into the history of timekeeping: from prehistoric humans to astronauts, these pages tell a story of mankind's 'invention' of time that begins with us looking at the moon and ends with us walking on it.

Those who surrender to the appeal of clocks and watches enter an endlessly engrossing world: a mechanical microcosm of components capable of the humble task of telling the time or the lofty one of predicting the movements of the stars. The fascination is the same for all – whether Pharaoh, eighteenth-century French queen, twentieth-century tycoon, or, in a much more modest way, me, when I was growing up in the 1970s.

At that time, battery-powered watches were all the rage, and old mechanical watches could be had for pennies in junk shops and jumble sales. I wore them until they broke or until I found another I liked. They were everyday objects, and yet I felt they possessed a beauty that, when on my wrist, ever so slightly improved my experience of life. They also amazed me in the way that they compressed their untiring functionality into a space about the size of a coin. Their dials boasted about the number of jewels they contained, their 'automatic' self-sufficiency, and their pride in being 'Swiss Made'. While their case backs proclaimed their impregnability with a list of attributes: waterproof, shockproof, dustproof, anti-magnetic…

I succumbed to their magic and have been under their spell ever since.

Just how many watches I accumulated became clear a couple of years ago, when my younger son retrieved a carrier bag full of old watches from the shed in the few square feet of open space behind our house that we ironically refer to as the garden. What amazed me was how many of them he coaxed back to life by merely applying thumb and forefinger and winding them up. He was able to enjoy them just as much as I once had, and with this rediscovered haul we recreated our own budget version of the Patek Philippe advertisement that reminds you that you never really own a

Left: Wheel mechanism of the clock at Strasbourg Cathedral (see pages 82–91).

Following pages: The fabled Graves supercomplication by Patek Philippe (see pages 198–209), the most complicated portable personal timepiece made before the use of computer-aided design.

watch, but merely look after it for the next generation (or, in this case, put it in the shed and forget about it for twenty years until the next generation comes across it by accident).

Sadly, my early excursions into watch-buying were spectacularly uninformed, so there were no Patek Philippes stashed in the shed. The reason that this advertisement has been so successful and has become so familiar, even to those who will never own – sorry, look after – a Patek, is that, while the time of day is freely available wherever we look, we continue to associate timepieces with an intrinsic value. (Why else do we clothe them in gold and jewels like a medieval reliquary?)

If nothing else, they are travellers from another time. I can never hear the chimes of the clock tower at the Palace of Westminster, whose stairs I climbed while researching this book, without thinking of the Victorians who built what we slightly erroneously know as Big Ben and placed it literally and figuratively at the heart of the world's largest empire. The empire is long gone, but the clock remains a visual shorthand for an entire nation and, along with the Eiffel Tower, Empire State Building, Taj Mahal, Colosseum and Great Wall of China, is one of the most famous structures ever built by the hand of man.

There is some unseen alchemy that transforms these mechanical objects into vessels of emotion. Sometimes those emotions intensify to such a pitch that a timepiece can inflame the passions of people willing to part with millions (almost £18 million in one memorable case) for the privilege of being its custodian and the chance to write themselves into the object's history.

It is their stories that make the timepieces in this book (and many others beside) such compelling objects, and I hope that those I have chosen will adequately convey the sense of wonder to be derived from such instruments – even from the, quite frankly, rather grubby 'Ishango' bone, of Palaeolithic Africa, where my loosely strung necklace of stories begins.

The Bare Bones *of* Time
Ishango Bone – Palaeolithic Era

A slender, curving, dirty-brown baboon fibula is not the most remarkable-looking item in the museum of Natural Sciences in Brussels, but it may just be the most important.

In 1950, 30-year-old geologist Jean de Heinzelin de Braucourt was deputy director of the laboratories of the Royal Belgian Institute for Natural Sciences. Already a distinguished academic, his lectures were some of the liveliest and best attended at the universities of Ghent. However, he was never happier than when out in the field, and on 25 April he arrived at Ishango, a few kilometres south of the equator on the northern shores of Lake Edward, at the mouth of the Semliki River, where elephant and hippopotami come to bathe. Today, Ishango is in the Democratic Republic of the

A familiar image from prehistory: a rock painting by the San people of South Africa's Cederberg Mountains, depicting an archer ready for the kill, with bowstring taut.

Congo, near the border with Uganda. In 1950, it was still the Belgian Congo, and Ishango was in the Albert National Park – 3000 square miles of lush, primal equatorial territory unchanged for thousands of years.

It might have been another routine expedition of late colonial-era exploration. Deep in the heart of Africa, Lake Edward had only become known to Western science in 1888, when it was recorded by Henry Morton Stanley and named after the then Prince of Wales. In 1925, King Albert of the Belgians declared the area a national park. The Royal Belgian Institute for the Natural Sciences had been sending expeditions to explore the park since the 1930s, and would continue to do so until the Congo became independent in 1960.

However, the expedition of 1950 would write the name of the remote ranger station of Ishango into the history of mankind with the discovery, by Heinzelin de Braucourt, of a baboon fibula, buried under layers of volcanic ash. This particular length of baboon bone had been fashioned into a tool, topped with a sharp piece of quartz. Yet, it was not the quartz that intrigued the young explorer – rather the three columns of grooves that had been painstakingly carved into it.

Time-stained and weathered with the passage of millennia, it was clearly old. Heinzelin de Braucourt reckoned it to be between 6–9000 years old; even in the late 1970s it was still judged to have originated from 6,500 to 8,500 years ago. Only later would it be discovered that it had actually been made by humans who had lived here among the apes and elephants during the Palaeolithic era, sometime between 20,000 and 25,000 BCE, which means the glyphs predate the written word.

Resembling a three-dimensional barcode, these comb-like groups of notches are a palimpsest upon which generations of scientists have written their own interpretations. At the moment, the weight of opinion favours the interpretation that the Ishango bone is some sort of tally stick, a prehistoric slide rule or calculator. To the layman's eye, the foundations upon which a significant scientific theory can be built can be as slender as… well… an old bone. More enthusiastic readings of these notches have evoked a picture of early mankind sitting about on the shores of a prehistoric lake pondering the mystery of prime numbers, even though one might have thought they had more pressing things on their minds, given the notoriously predatory fauna with which they shared the continent.

Another school of thought would have us believe that these notches are so old that they may, quite literally, be as old as time itself: specifically, that the Ishango bone is the world's oldest timepiece.

Alexander Marshack was a *LIFE* magazine writer and photographer turned popular scientist. Notwithstanding the absence of a row of letters after his name proclaiming his academic credentials, he was associated with the Peabody Museum of Archaeology and Ethnology at Harvard University.

He had learned of the Ishango bone in the 1960s, when Heinzelin de Braucourt wrote up his experiences in the *Scientific American*. At the time he was engaged in writing the historical section of a book about the scientific development that had led to the age of lunar exploration. And while researching the history of human development, he kept returning to the question of the calendar. Calendrical understanding in ancient civilisations was accepted, but after contacting scholars and experts he repeatedly discovered that such calendars were already highly developed, and rather than this knowledge having arrived all of a sudden, it seemed logical to him that they had been preceded by something simpler.

It was while studying pictures of the Ishango bone that he had a moment of almost Archimedean insight. It occurred to him that the notches on the bone might be more than idle scratches or hunting tallies, but instead some sort of very early lunar

Alexander Marshack (seated) developed theories that revolutionized understanding of engraved bone fragments as calendrical devices.

Jean de Heinzelin de Braucourt, in the wardrobe of the prototypical late Colonial explorer, on the bank of Lake Edward in 1950.

calendar. The way he interpreted them, the marks were consistent with a lunar cycle of two months. 'I was dizzied,' he later told the *New Yorker*. 'It seemed too easy and I distrusted it.'[1]

With the backing of the National Geographic Society, the National Science Foundation and various private bodies, he spent the 1960s researching, analysing and interpreting thousands of Palaeolithic carved and notched objects. So seriously was his work taken that, even at the height of the Cold War, he was allowed access to collections in the Soviet Union.

As he investigated, his hunches about the Ishango bone grew into a temporally centred hypothesis about human development. Calling his approach 'cognitive archaeology',[2] he speculated that it was with an understanding of time and sequential thought that the journey from hominid life in Palaeolithic Africa, to what we know as human civilization today, began.

His theory required a new understanding of early man. Instead of viewing his differentiating characteristic as being able to conceive, fashion and use tools, Marshack's man owed his rise to an understanding of time – as well as his ability,

An engraved bone that could be an early ancestor of the pocket watch.

A few kilometres south of the equator on the northern shores of Lake Edward, at the mouth of the Semliki River, Ishango is where elephant and hippopotami come to bathe. Part of a 3000-square-mile park of lush, primal equatorial territory unchanged for thousands of years, it seems an unlikely place to begin the history of the timepiece.

however rudimentary, to track it. As a *New Yorker* article on Marshack explained in 1974, he believed that it was 'the capacity for observing, and for remembering for future use, the cyclical patterns of the Seasons and the plants, and of the animals, whose inability to think in time gave man such a powerful advantage over them. This slowly evolving tendency, he reasoned, was what had led to such human developments as toolmaking and spoken language.'[3]

Marshack's theory was that, in order to observe, understand and note the waxing and waning of the moon, then to interpret, communicate and apply this information, our early ancestors had a well-developed spoken language. For Marshack, scratched markings on bone were the precursors of writing and numeric notation – protean attempts to store and retrieve information.

If the word civilization is most broadly understood to imply a sense of order, then Marshack's theory makes a compelling case for an understanding of time being an important part of the foundation upon which human civilization was built. The link made between the passage of time and the recurrence of seasonal events would have had obvious importance for hunter-gatherers, who would be very interested

in forecasting the migratory patterns of their quarry. And, of course, thousands of years later, the move from hunting and gathering to an agricultural system would have been impossible without an appreciation of time: without it, how would one know the most propitious times to sow and harvest, or be able to calculate the quantity of food necessary to sustain the community in the interval between planting seeds and reaping the crop?

Marshack found Palaeolithic Europe particularly fecund soil for his theory that prehistoric man ordered his life into a temporal framework. He believed that serpentine groups of marks of varying thickness on stones or scraps of antler and bone were calendars that recorded the phases of the moon. Eminently useable on, say, a hunting expedition, thanks to their small size, they were, in effect, the first portable personal timepieces.

By 1972, he had worked this up into *The Roots of Civilization*, a lavish work of popular archaeology-science-history – a genre that was popular at the time. The late 1960s and early 1970s saw increasing interest in the origins of human civilization; at one end there was Erich von Däniken's 1968 bestseller *Chariots of the Gods*, which put forward a theory of alien intervention in mankind's development; while at the other was the BBC's rather more serious collaboration with Marshack's old employer, Time-Life, *The Ascent of Man*, a television series presented by mathematician and historian Jacob Bronowski, which aired in 1973.

The Ishango bone, found on an archaeological site on the Congolese bank of Lake Edward could be among the world's earliest time-reckoning devices.

The Roots of Civilization was thus perfectly aligned with the zeitgeist. Writing in the journal *Current Anthropology* in 1972, Marshack proposed a theory of widespread contemporaneous temporal awareness: 'The lunar hypothesis suggests that the Upper Paleolithic [*sic*] notations represent one class of symbol and therefore a specialized cultural usage of the general cognitive capacity. The evidence suggests that this capacity for sequential and periodic conceptualization was at a comparable stage of development among different groups in different environments and regions.'[4] To Marshack, it seemed that he had found evidence of a nascent understanding of time dawning at roughly the same time in different parts of the world.

Marshack's theories were deemed revolutionary and, of course, there was a counter revolution. 'In the 1990s some younger scholars took exception to Marshack's structuralist interpretations, preferring to see more magical and religious motives behind Palaeolithic phenomena and decrying Marshack's approach as excessively numerological,' observed his obituary in *The Times*, adding: 'While some of the caveats seem valid, the extent to which Marshack reorientated attitudes to the intellectual processes and achievements of our distant ancestors should not be underestimated.'[5]

Inevitably, at a distance of tens of thousands of years, how, why or when early humans started to mark the passage of time is uncertain. However, what *is* certain is that time moved more slowly two dozen millennia ago. A day was the smallest temporal unit, and the lunar calendar joined those days up, then matched them to a cycle of planetary activity – thus giving a representation of both the cyclical and progressive nature of time.

When it comes to the discovery at Ishango, whatever the objections, it is tempting to think that this unprepossessing simian bone is the ancestor of time as we read it today on wristwatches and many other places besides. That those Stone Age people crouching on the edge of an equatorial lake almost 30,000 years ago had already worked out that the measurement of time, whether in scintillas of a second or by the cycle of the moon and the passing of the seasons, is how we make sense of our lives.

1. *The New Yorker*, 22 April 1974
2. *Ibid.*
3. *Ibid.*
4. Alexander Marshack, 'Cognitive Aspects of Upper Paleolithic Engraving', *Current Anthropology*, Vol. 13, No. 3/4 (June–October 1972), pp. 445–77
5. *The Times*, 22 January 2005

Pit Stop
Warren Field Calendar – Mesolithic Era

Known to historians as Mesopotamia, the fertile arc of river-nourished land between the Mediterranean and the Persian Gulf is generally accepted as the place 'where the world's earliest civilization developed'.[1] Its clay bricks, clay tablets, clay figures, clay seals, cuneiform script and splendidly bearded monarchs are taken for granted as betokening the birthplace of human civilization.

But, according to a team of archaeologists led by Professor Vincent Gaffney of Birmingham University, it is not necessarily to the banks of the Tigris and Euphrates that we should be looking; rather to the land between the Rivers Dee and Don that spring from Scotland's Cairngorm mountains.

Not far from the north bank of the River Dee is Warren Field, which looks like just another non-descript stretch of agricultural land. However, when leading landscape archaeologist Professor Vincent Gaffney of the University of Bradford arrived in 2013, he saw something rather more significant.

As early as the long dry summer of 1976, aerial photography had detected evidence of a site worthy of archaeological investigation. Excavation in 2004, 2005 and 2006 had uncovered a series of twelve differently shaped pits arranged in a fifty-metre arc.

A Warren Field pit during excavation.

The Warren Field pit alignment recorded as cropmarking on an aerial photograph.

Given that the remains of a large Neolithic hall were nearby, it may have seemed reasonable to suppose that this find also dated from around 4000 BCE. Carbon dating, however, told a very different story. These pits were old – very, *very* old – and had been dug during the Mesolithic era, around 10,000 years ago.

It had clearly been an important monument of some sort, constructed over a period of a couple of centuries, and had remained in use until at least the beginning of the Neolithic period, some 4000 years later. For years, the purpose of these prehistoric pits remained a mystery. It was not until 2013, having applied the most advanced remote sensing technology, along with software that mapped the sunsets and sunrises as they would have appeared over this landscape at that period in history, that Professor Gaffney put forward quite a startling hypothesis.

'The pits had been studied vertically,' he stressed, but according to Gaffney, it was in their *alignment* that their real significance lay. 'When you look at the alignment it points to the Slug Road Pass, the major pass south of the Dee Valley.' Next came the complicated part: finding out where the sun would have appeared in 8000 BCE. However, before that could happen, one of Professor Gaffney's colleagues had to write some new software, since the programme ordinarily used to give the apparent position of the sun vis-à-vis local landmarks did not work for periods earlier than 4000 BCE. This was because such monuments marking solstices were thought to customarily date from the Neolithic period, of which Stonehenge is one of the most famous, while the

According to Professor Vincent Gaffney of Birmingham University, the land between the Rivers Dee and Don that spring from Scotland's Cairngorm mountains is as important as Mesopotamia in the history of time reckoning.

Goseck Circle in Germany is one of the oldest, dating from around 4800 BCE. The Warren Field pits were thousands of years older than both of these.

The results of this modelling were remarkable: around ten millennia ago, sunrise on the winter solstice would have occurred directly over the Slug Road Pass. 'You can imagine someone walking through the leafless forests at dawn,' Gaffney says, 'observing the sun coming through the Slug Pass and rolling along the valley – it would have been an impressive sight.'

He goes on: 'The sunrise along that valley has an astronomic significance; and the only thing that the strange shape of the pits reminds us of is the phases of the moon. If you look at most early society, lunar observation tends to be the beginning of time reckoning, as it is the only heavenly body movement that is not annual: the phases happen within a month.'[2]

Shaped to represent the various phases of the moon, with a two-metre-wide pit for the full moon in the centre, the Warren Field pits displayed twelve lunar months. Yet it was only by viewing the pits in their topographical context that their full use became clear: 'A group of hunter-gatherers who had presumably been observing the skies and celestial bodies for hundreds of years took the decision to build a monument [with] which they could track the time month by month using observations of the moon.'[3]

However, as we saw in the Introduction, the solar year does not coincide exactly with the lunar calendar of twelve months; the lunar year lasts about 354 days, whereas a solar year comprises a little less than 365.25. It would not take long for a solely lunar-based time system to become useless, therefore, and so in order to remain relevant to the changes in season, and to anticipate seasonal events, the pits would need to be 'reset' every year. Millennia later, extra days would be inserted into calendar years to arrest this drift, but, unlettered and innumerate as he was, Mesolithic man had not yet reached that level of calendrical refinement.

A virtual model showing the midwinter solstice viewed from a pit calendar.

In the case of the Warren Field pits, this recalibration was easy given the alignment with the Slug Road pass, over which the sun rose at dawn on the midwinter solstice, permitting the pits' users to start the lunar cycle all over again. 'It seems to indicate the point at which hunter-gatherers have the capacity or the need to actually have a formal approach to time itself and in doing that they are not just thinking of what happened in the past, they are anticipating the time to come and they are probably scheduling activities as a consequence of that and that can lead to all sorts of social change.'[4]

In a world with few distractions, after dark there was little to do but watch the night sky. Observation of the heavens could, argues Gaffney, lead to development of a belief system founded upon the observed behaviour of heavenly bodies, which in turn would reinforce a link between the moon and the sun and events on Earth. Given the behaviour of tidal waters, this is not nearly as primitive as it at first may sound.

If early artefacts, such as the Ishango bone, were ways of tracking time in order to be in the right place at the right time, for animal migrations for example, then the River Dee was one of the right places for the hunter-gatherer to be when the salmon swam up the river. According to Gaffney, this annually recurring riparian congregation would acquire a social significance greater than merely meeting nutritional needs. 'In a society which is largely mobile, this would see large groups come together and carry out social and religious rituals on the Dee.'[5]

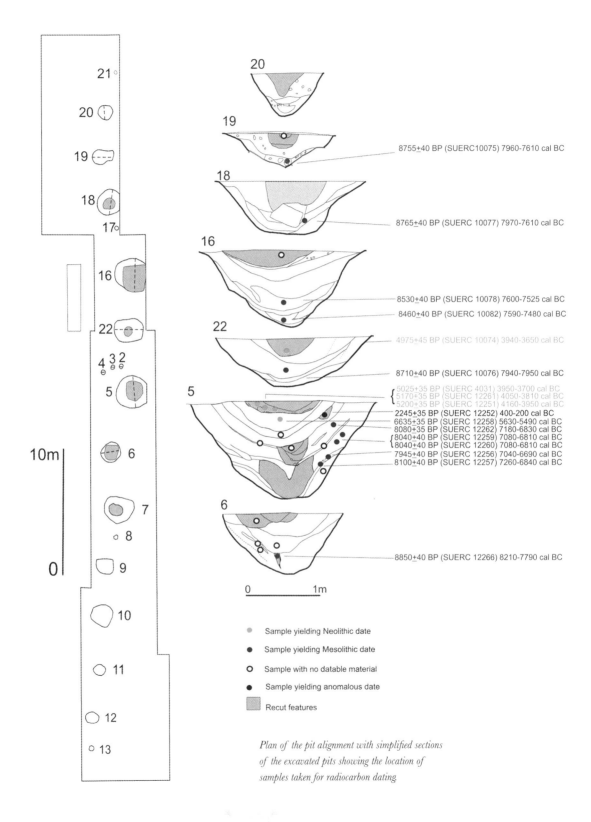

8755±40 BP (SUERC10075) 7960-7610 cal BC

8765±40 BP (SUERC 10077) 7970-7610 cal BC

8530±40 BP (SUERC 10078) 7600-7525 cal BC

8460±40 BP (SUERC 10082) 7590-7480 cal BC

4975±45 BP (SUERC 10074) 3940-3650 cal BC

8710±40 BP (SUERC 10076) 7940-7950 cal BC

5025±35 BP (SUERC 4031) 3950-3700 cal BC
5170±35 BP (SUERC 12261) 4050-3810 cal BC
5200±35 BP (SUERC 12251) 4160-3950 cal BC
2245±35 BP (SUERC 12252) 400-200 cal BC
6635±35 BP (SUERC 12258) 5630-5490 cal BC
8080±35 BP (SUERC 12262) 7180-6830 cal BC
8040±40 BP (SUERC 12259) 7080-6810 cal BC
8040±40 BP (SUERC 12260) 7080-6810 cal BC
7945±40 BP (SUERC 12256) 7040-6690 cal BC
8100±40 BP (SUERC 12257) 7260-6840 cal BC

8850±40 BP (SUERC 12266) 8210-7790 cal BC

0   1m

- Sample yielding Neolithic date
- Sample yielding Mesolithic date
○ Sample with no datable material
● Sample yielding anomalous date
▨ Recut features

Plan of the pit alignment with simplified sections of the excavated pits showing the location of samples taken for radiocarbon dating.

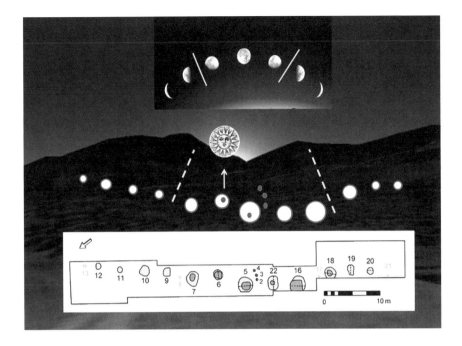

The plan of the Warren Field pit alignment below the symbolic arrangement of the pits in relation to the Slug Road pass. (The backdrop has been exaggerated for display purposes.)

Instead of a calendar, Gaffney describes the pits as a 'time reckoner', predicting the future and ordering the past. Whatever it is called, however, it indicates advanced cognitive development: 'Time is a social construct. It does not, as such, exist; things happen, but that is all.'[6] And with the development of the abstract concept of time over several hundred years, observing first the annual solar event and then the monthly lunar events, so the very first glimmers of the dawn of civilisation can be detected.

As Professor Gaffney puts it: 'What we are looking at here is a very important step in humanity's earliest formal construction of time, even the start of history itself.'[7]

1. Britannica.com
2. Interview with the author, October 2018
3. BBC News, 15 July 2013, Vincent Gaffney interviewed by Huw Edwards
4. *Ibid.*
5. Interview with the author, October 2018
6. *Ibid.*
7. Roff Smith, 'World's Oldest Calendar Discovered in UK', *National Geographic*, 16 July 2013 (https://news.nationalgeographic.com/news/2013/07/130715-worlds-oldest-calendar-lunar-cycle-pits-mesolithic-scotland/)

A Bucket *with a* Hole
The Karnak Clepsydra

The accidents of history, grave robbery and archaeology have decreed that the most famous of Egyptian pharaohs was the boy-king Tutankhamun. However, it was under the rule of his grandfather, Amenhotep III, that Egypt, stretching from the Euphrates to the Sudan, reached the zenith of its glory, power and brilliance.

Amenhotep III is remembered for one of the most remarkable building sprees of antiquity. His mortuary temple was 'the grandest of all mortuary temple complexes built in Egypt',[1] he created an entirely new place of worship at Luxor and embellished the sprawling 200-acre temple complex of Karnak. He also revived the cult of the sun god Ra and promoted worship of the sun disc Aten – even taking the sobriquet 'Dazzling Aten' for himself. He was a sun king millennia before Louis XIV – and Karnak was his Versailles.

And yet, by the time French Egyptologist Georges Legrain looked on the home of Dazzling Aten in the late nineteenth century, it recalled the lines of Shelley's *Ozymandias*. It was a ruin; precarious needles of stone, vast tumbled columns and jumbled piles of huge hewn stones dwarfed the European visitors, students of the

The ancient ruins of the sprawling Karnak temple complex in Luxor, Egypt.

new science of archaeology, in their broadbrimmed white hats, three-piece suits, high collars and bow ties. Legrain had studied at the Beaux-Arts in Paris and attended lectures on Egyptology at the Sorbonne. In 1895 he was named Director of Works at Karnak. From then until his death in 1917, he made Karnak his life. A passionate photographer, he was also able to capture the enormity of his task in all its gigantic decay.

Legrain's discoveries have entered the annals of archaeology. The finding of the so-called Cachette of Karnak in 1903 yielded one of the richest troves of the arts of antiquity: for four years, hampered by a high water table, Legrain and his workers hauled out over 700 statues, 17,000 bronzes and sundry other objects. His single most spectacular find came in 1907, however: a huge pink granite scarab atop its own column, discovered near the Sacred Lake.

Legrain himself became synonymous with Karnak and was a celebrity of sorts, welcoming other celebrities who made the pilgrimage down the Nile and into the past. Like modern visitors they posed for souvenir photographs, but dressed more formally. Today there is something slightly comical about a bearded and boatered Camille Saint-Saëns being shown the pink granite scarab by a dark-suited, pith-helmeted Legrain.

Amenhotep III, the Dazzling Aten, is remembered for one of the most remarkable building sprees of antiquity. During his reign, the king presented offerings to a statue of himself as a manifestation of the sun god. This image depicts the royal statue and the sled (c. 1375 BCE) on which it was transported for these ceremonies.

Theirs were the golden days of archaeological discovery, when Egyptomania swept the European imagination up in its embrace, influencing every aspect of Western culture from cinema and detective fiction to interior design and jewellery.

Amidst all the excitement, it would have been easy to overlook the discovery of a small alabaster flowerpot, 34.6 cm in height, with a diameter narrowing from 48 cm to 26 cm, in what has been described as a refuse tip near the site of the temple of Amon.

Compared to such photogenic show-stealing finds as the giant scarab, this vessel is attractive rather than evocative. Seen against Karnak's towering pillars and imposing statues, pregnant with the mysteries of an ancient, half-known civilization, this simple survivor of life in the time of Amenhotep III appears comparatively humble. Nevertheless, it is clearly a thing of quality: etched into the pale, translucent material were three bands of decorative bas-reliefs depicting animals, deities, mythical beings and actual historical figures, including Amenhotep III himself.

Ironically for a man who identified himself with the sun, it is from his reign that the first evidence of man's quest to liberate time from the sun survives. Egyptian and Babylonian society told the time using sundials, which, although often quite sophisticated, were ultimately dependent on the presence of sunlight – which is what makes the 'flowerpot' found at Karnak so fascinating.

Its embellished, tapering form becomes hugely significant when one notices the tiny hole in the bottom. This 'flowerpot' or 'bucket' was in fact a clock, and as the water dribbled out of the bottom, the time could be read on an interior scale by the level of the remaining water (the tapering sides compensated for the effects of the change in water pressure as it emptied). The exterior, richly carved with hieroglyphics and representations of constellations and deities, showed, at its top, gods and the thirty-six 'decan' stars that rose one after another at night and were used by the Egyptians as a star clock. At the bottom was a calendar of months.

Reconstruction of an early Egyptian water clock, 1415–1380 BCE, the 'flowerpot' that told the time was found at Karnak Temple, Upper Egypt, in 1904, and dates from the reign of King Amenhotep III (1415–1380 BCE). In use, the vessel was filled with water, which leaked out slowly from a small hole near the bottom; the time being indicated by the level of the water remaining within.

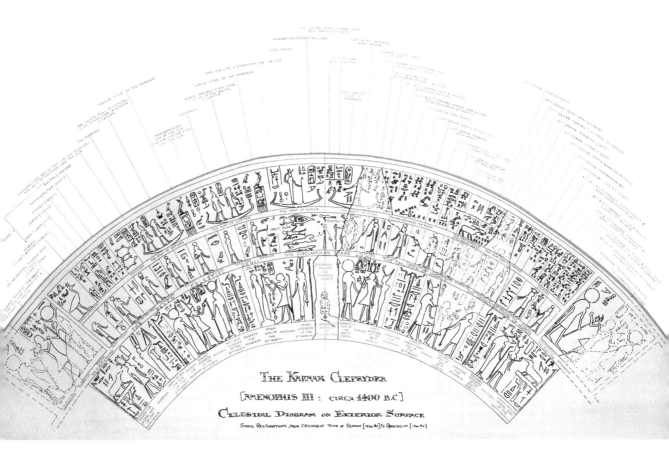

The Karnak Clepsydra, Egyptian celestial diagram, c. 1400 BCE. A copy (from 1939) of the hieroglyphics and other figures embossed on the outside of the water clock or 'clepsydra' which was discovered in the Temple of Karnak, Luxor, Upper Egypt. The top row shows a series of planet gods and the thirty-six decan stars – the great celestial timekeepers of the ancient Egyptians. In the middle are various constellations and deities and, on the bottom row, a calendar of months and month gods.

Literally translated as 'water thief', a clepsydra uses the dropping level of water in a calibrated vessel to mark the passage of time. This was the world's first (surviving) accurate timepiece, capable of functioning day and night, and it would give mankind the time for the next 3000 years. It was a way of regulating life and a powerful symbol of how advanced human civilization had become. A calendar that could predict seasonal climatic events based upon the regular appearance of new moons and the annual solar cycle was sufficient for an agricultural society, but the clepsydra was different.

This water clock betokened a society that lived a life so complex and involved that merely gazing at the sun to judge the approximate 'hour' of the day was no longer sufficient: religious ceremonies needed to be conducted, the business of government

Suitably dressed for the heat: Camille Saint-Saens and Georges Legrain admire the giant granite scarab at Karnak temple.

needed to be carried out, the hundreds of tiny administrative tasks that together make civilization needed to be coordinated, and for all that, an accurate idea of time was needed. For that reason, the clepsydra would outlive 'Dazzling Aten' and remain in use long after the sands of the desert had started to reclaim the proud structures of Ancient Egypt.

Karnak was a must-see on the itineraries of late nineteenth and early twentieth century travellers. This photograph, taken in January 1877, shows American financier Pierpont Morgan and his family picnicking at Karnak (from Herbert L. Satterlee's J. Pierpont Morgan, An Intimate Portrait, *1939).*

Aristotle records how the clepsydra was used to impose a time limit upon pleas in court (they were later used for the same purpose in republican Rome). And it was in Alexandria, the city founded by Aristotle's star pupil Alexander the Great, that the clepsydra achieved the status of a domestic luxury, a status-affirming object. The Alexandrian mathematician Ctesibius designed a clepsydra of such exigence that only in the seventeenth century would mankind devise a more accurate clock. Later, Hero of Alexandria would create other hydro-mechanical devices, such as a water clock to regulate a wine dispenser, simply to divert the rich and idle patrons for whom they were made. Thousands of years later, worshippers of religions that did not even exist in Karnak's heyday would be called to prayer by clepsydrae.

And it all began with a bucket with a hole.

---

1. World Monuments Fund (https://www.wmf.org/project/mortuary-temple-amenhotep-iii)

Following pages, Left: Reconstruction of a clepsydra (water clock), invented by Ctesibius of Alexandria, c. 270 BCE, an inventor and mathematician who initiated a tradition of great engineers in ancient Alexandria. This clock worked by water dripping at a constant rate and raising a float with a pointer.

Following pages, Right: A clepsydra that measures time by how long it takes a vase to fill with water, acting on a dial as the water rises. Drawing by an unnamed artist for Morris's series of cigarette cards 'Measurement of Time' (1924).

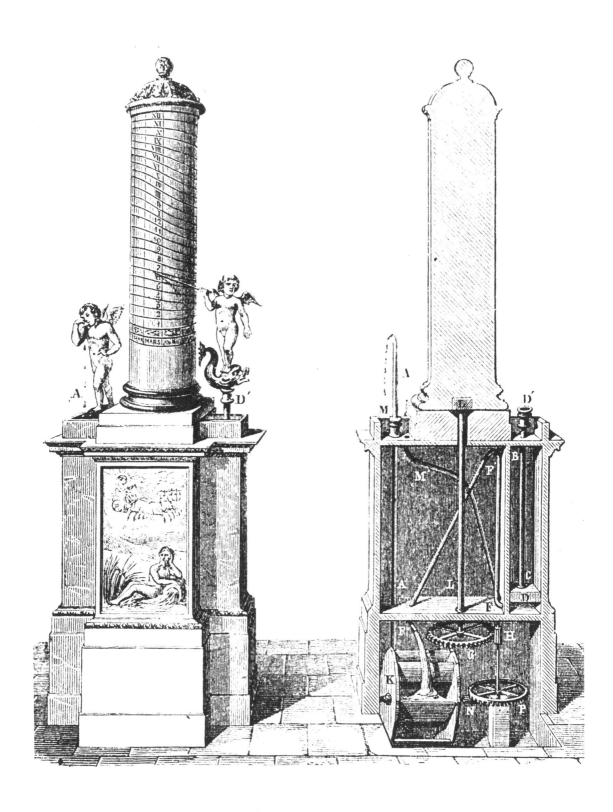

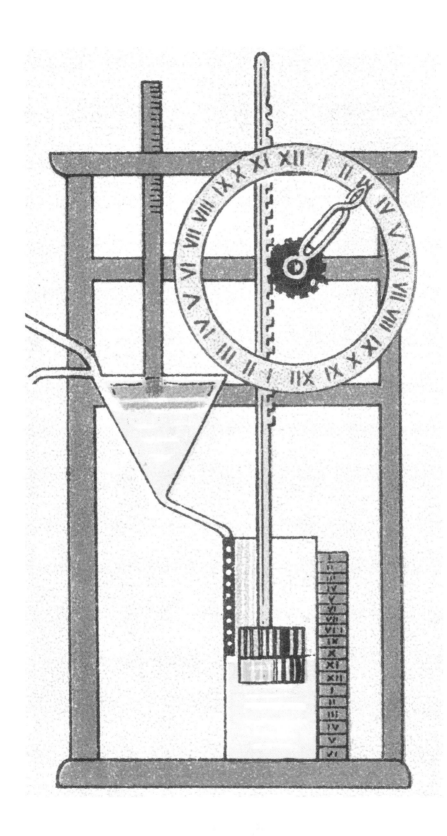

Back *to the* Future
The Antikythera Mechanism

The island of Antikythera is a sparsely inhabited outcrop of rock that juts above the waves, about half way between the Peloponnese and the island of Crete. It is the sort of place where even today the wild goats outnumber the few dozen human inhabitants.

However, to Captain Dimitrios Kondos, seldom had land looked more inviting. The two caïques he commanded had been blown off course and, in search of shelter in which to wait out the storm, he dropped anchor off Antikythera. It was Easter 1900, and Captain Kondos was leading a sponge-fishing expedition of six divers and twenty-two oarsmen who were returning home from the seas off Tunisia to the island of Symi.

The main fragment of the Antikythera machine.

The choppy waters and rocky coast of Antikythera.

Sponge fishing was to the Dodecanese islands what whaling had been to Melville's Nantucket. In the closing decades of the nineteenth century, new technology had revolutionised an activity little changed since antiquity. The introduction of canvas and rubber diving suits with solid spherical helmets liberated divers from the restrictions imposed by the capacity of their lungs. And though the modern methods brought the new peril of decompression sickness, they also brought new wealth.

After the storm had abated and the seas had calmed, Kondos' team of divers decided to see if there were any sponges to be added to their haul. Elias Stadiatis soon returned and told his captain that he had seen a pile of dead bodies on the seabed.

Astounded, Captain Kondos decided to take a look for himself. At a depth of forty-two metres, he found the wreck of large ancient ship surrounded by the jumbled amphorae that had spilled from the vessel and that would become a familiar sight for marine archaeologists working in the Aegean and Mediterranean during the rest of the century. But this was more than a jobbing cargo ship transporting wine and olive oil between the ports of the Hellenistic eastern Mediterranean: 'The real excitement… was not so much in the ship itself but in a treasure that was plainly visible – a pile of bronze and marble statues and other objects.'[1]

One such object was the right arm of a bronze classical sculpture. Made to a scale larger than life, encrusted with detritus and discoloured after centuries underwater, it was removed from its resting place and taken aboard Captain Kondos' caïque.

A reconstruction of the Antikythera Mechanism on display at the exhibition of Ancient Greek Technology in Athens, part of the second international Conference on the Ancient Technology (2005). Comprising a minimum twenty-nine gears of various sizes that were moved simultaneously via a handle: its discovery rewrote the history of astronomy.

Back on his boat, he took bearings so that he would be able to locate the site again, and then sailed for home.

Captain Kondos and his crew had just discovered an important underwater archaeological site, far more interesting and valuable than any sponge, but its significance seems to have eluded them, or was eclipsed by more immediate concerns. One account of the conclusion of this voyage says the crew 'spent some six months in the riotous living that was customary on completion of a successful trip'.[2]

Riotous though the living may have been, the decision they took about the wreck was a sober one. 'Deciding to approach the correct authorities, rather than make it an illicit private adventure, Kondos and Stadiatis went to Athens, taking the bronze arm with them.'[3]

The authorities and the sponge fishermen reached a financial arrangement and exploration of the site was undertaken. Contemporary photographs show its fruits stacked neatly in the museum's store; some clearly parts of sculptures awaiting reassembly, others calcified, eroded and looking a little like the featureless but nonetheless human forms found during excavations of Pompeii. The Antikythera Youth, the name given to the larger-than-life-size bronze salvaged from the wreck, is now regarded as one of the masterpieces of antiquity.

Less readily identifiable artefacts were brought to the surface too, among them dozens of pieces of corroded bronze covered with indistinct inscriptions and the outlines of what resembled toothed wheels. For all the lithe athletic grace and muscular beauty of the Antikythera Youth, it could easily be argued, however, that these shards of metal, at which the sea had eaten for 2000 years, were of far greater significance – and yet it was not until 1902 that they were first inspected.

The front side of the reconstruction of the Antikythera Mechanism. The Antikythera Mechanism now has its own gallery in which all of its fragments are on display.

Initially scholars were simply divided as to whether these lumps had once been an astrolabe or, as others posited, something more complex. Over the coming decades, academics weighed in on the debate, but progress on discovering what the so-called Antikythera Mechanism actually did and how it functioned was slow. Then, in the middle of the century, a new figure entered the debate.

'From about 1951, in the course of investigations into the history of scientific instruments with special reference to ancient astrolabes and planetaria, I began to appreciate the deep significance of the Antikythera mechanism,'[4] wrote Professor Derek de Solla Price during the 1970s, following more than twenty years examining and pondering this object. He had first heard of it when, as a bespectacled, pipe-smoking young academic, he was working on a second PhD, on the history of science, at Cambridge. De Solla Price had first made his name when he had come across a sixteenth-century manuscript on a sixteenth-century astronomical device at Peterhouse, Cambridge. At least it had long been *thought* to be of sixteenth-century origin. In fact, De Solla Price established that it was two centuries older and reattributed it to the most famous English author of the Middle Ages: Geoffrey Chaucer. Although more widely known as the author of *The Canterbury Tales*, thanks to this sleuthing young academic he was now also recognized as a writer on astronomy and author of *A Treatise on the Astrolabe*.

However, the re-attribution and the re-dating of a major medieval text was but an *amuse bouche* for the banquet of academic controversy that he was about to serve up when he turned his attention to the Antikythera Mechanism, which he knew only from old photographs. He was fascinated by the highly advanced gearing that he spotted and, in 1953, he received modern images from Greece, taken after the device had been cleaned. These showed even more intriguing details and led him to propose a theory amply encapsulated in the oxymoronic title of his 1955 paper, 'Clockwork before the Clock'.

The recovery of the items from the Antikythera wreck marked the beginning of modern undersea archaeology.

It was not until 1958 that he actually travelled to Athens to examine the fragments themselves. What he discovered caused him to upgrade his initial theory to one that he promulgated in the *Scientific American* a year later with the article 'An Ancient Greek Computer'. The headline, especially in such a respected publication, was more provocative than scholarly. Moreover, in 1959, the computer was still a largely experimental device, about the size of a room or a wardrobe, with blinking lights and whirring tape drives. Just how could a 'primitive' society have created anything similar 2000 years earlier? Unsurprisingly, the Antikythera debate spilled out from academic journals and into the public arena.

By then, De Solla Price was working in America, first at Princeton, then Yale, and the resulting controversy took him by surprise. 'I must confess that many times in the course of these investigations I have awakened in the night and wondered whether there was some way round the evidence of the texts, the epigraphy, the style of construction and the astronomical content, all of which point very firmly to the first century BC.'[5]

Diving at the site of the Antikythera wreck at the beginning of the twentieth century.

Some thought that he had been fooled by the appearance of age and that in fact the mechanism was of nineteenth-century origin. 'Then again there were some only too ready to believe that the complexity of the device and its mechanical sophistication put it so far beyond the scope of Hellenistic technology that it could only have been designed and created by alien astronauts coming from outer space and visiting our civilization.'[6]

As mentioned earlier, the period in which De Solla Price was making these observations abounded with a popular curiosity about the origins of civilization, giving rise to new theories like Alexander Marshack's reinterpretation of Palaeolithic artefacts. At the same time, space exploration was a dominant cultural trope that found itself expressed in everything from the foundation of NASA to the tail design of American cars, so its intrusion into Hellenistic scholarship was not entirely unexpected. Sadly, De Solla Price did not adhere to the belief that the mechanism was of extra-terrestrial origin. His view was that previous generations of historians had simply got it wrong and had underestimated the scientific capabilities of the Greeks at that time: 'a drastic underestimation that can now be corrected.'[7]

De Solla Price was a man on a mission, and having spent the 1960s pondering these chunks of corroded metal, during the next decade he came up with the idea of using gamma-radiography to 'see through the corrosion'.[8]

He had wanted to use X-rays, but there had been difficulties getting 'heavy electrical power within the museum'.[9] The gamma rays gave him hope and he went, as anyone in his situation would have done, straight to the Greek Atomic Energy Commission, whose members, rather than sending the mad, pipe-smoking professor packing, listened attentively and said they would be only too happy to help. What the gamma rays revealed was 'a calendrical Sun and Moon computing mechanism'[10] of immense sophistication involving 'more than thirty gear-wheels'[11]. It was a breakthrough in the understanding of Hellenistic culture that, in his opinion, placed the Antikythera Mechanism at the beginning of a direct line of succession that led to the industrial revolution.

As it happens, even De Solla Price, the man who had inspired some to believe that this was a piece of alien technology, had slightly underestimated the device's capabilities. As Yannis Bitsikakis, curator of the Antikythera Shipwreck exhibition at the National Archaeological Museum in Athens in 2012, explains, 'his model was performing simple operations in a too complex way', i.e. the mechanism was capable of doing far more than he thought.

Derek de Solla Price (1922–1983). Behind the genial pipe-smoking exterior lay the brilliant mind of a gifted and daring horological scholar, unafraid to challenge received wisdom.

De Solla Price died in the early 1980s, but the interest in the object that he had spent a long period of his life investigating continued. From 1990 onwards, the fragments were subjected to further photography, tomography, X-rays, radiography, as well as inspection by physicists and astrophysicists. From 2005, they were examined with advanced surface imaging and high-resolution three-dimensional X-ray tomography, performed by a specially designed twelve-ton X-ray scanner (named Blade Runner) that had been created expressly to unlock the secrets of the eighty-two Antikythera fragments that, Bitsikakis says, could be carried in a shoe box. It was a high-tech sledgehammer designed to crack an ancient technological nut.

What twenty-first-century science finally revealed surpassed even De Solla Price's hypothesis, and at the same time proved conclusively that it was not a device designed by visitors from outer space – unless they were keen sports fans. Blade Runner revealed unknown inscriptions and 'a 4-year "Olympiad" dial which was tracking the sequence of the ancient Panhellenic games' [12].

The ineffable beauty and grace of the Antikythera Ephebe – bronze sculpture, 350 BCE (194cm) at the National Archaeological Museum, Athens.

Conceptually, it linked the activities of man to the heavens, placing him at the centre of cosmic events that he was able to predict. As well as the Panhellenic games, the Metonic calendar was also represented. 'By turning a handle, pointers moved around dials and displayed the positions of the Sun and the Moon over the zodiac scale and the solar calendar, the phases of the Moon, the year and the month on the nineteen-year luni-solar calendar, and the possibility of lunar and solar eclipses on a 223-month period. Subsequent research showed that the positions of the five planets known in antiquity were most probably also displayed. So this Mechanism was a miniaturized "Cosmos", linking the human society and its time-keeping calendars to the ancient Universe.' [12]

The mechanical replication of the movements of heavenly bodies above him offered mankind the precious illusion of some sort of control over his existence: predicting astronomical events began to demystify human existence and the wider world – the sort of feat previously accomplishable only by oracles and gods.

This mechanical marvel that spent twenty centuries at the bottom of the Aegean was the successor of the clepsydra, the water-powered timekeeper that divided day and night into hours, and the predecessor of the mechanical clock. Recreating the journeys of the planets and stars through the celestial vault brought man a step further away from the beasts and a step closer to the divine.

1. Derek de Solla Price, 'Gears from the Greeks. The Antikythera Mechanism – A Calendar Computer from CA. 80 BC', *Transactions of the American Philosophical Society*, Vol. 64, No. 7 (1974), pp. 1–70
2. *Ibid.*
3. *Ibid.*
4. *Ibid.*
5. *Ibid.*
6. *Ibid.*
7. *Ibid.*
8. *Ibid.*
9. *Ibid.*
10. *Ibid.*
11. *Ibid.*
12. Yannis Bitsikakis, 'On Time', *Vanity Fair*, Autumn 2012

The Ides *of* March
Julian Calendar

*K*een *students of the Monty Python ouvre can doubtless recite the entire 'what have the Romans ever done for us' sketch, one of the most enduring cultural legacies of the exercise in comic blasphemy that is the 1979 film* Life of Brian.

For those unfamiliar with the film, John Cleese plays 'Reg', a first-century rebel leader rousing his rabble with some anti-Roman rhetoric. Having established that Romans have bled his people white for generations, Reg reaches his rhetorical crescendo with the question: 'And what have they ever given us in return?!'

Julius Caesar 100 BCE– 44 BCE. From Imperatorum omnium orientalium et occidentalium verissimae imagines ex antiquis numismatic... Addita cuiusque vitae description ex thesauro Iacobi Stradae. *A 1559 woodcut by Andreas Gesner.*

44

Calendar with the months from July to December, 25 AD, found at the archaeological site of Amiternun, near L'Aquila, Abruzzo, Italy.

'The aqueduct,' comes the simple answer, which is then followed by a flood of suggestions that ends with Cleese's question sounding altogether less snappy: 'All right, but apart from the sanitation, the medicine, education, wine, public order, irrigation, roads, the fresh water system, and public health, what have the Romans ever done for us?'

To which one might have answered: 'Sorted out the calendar.'

By 46 BCE, it was generally acknowledged that the calendar was a mess. The lunar cycle may have provided the concept of the month, but it was the solar cycle that determined the seasons, and reconciling the two was a problem with which mankind had been wrestling since the days of the Warren Field pits.

Based on the Greek lunar calendar, a year in Ancient Rome comprised ten months of thirty or thirty-one days, leaving fifty or so days to make up. The solution was to stop the year at the end of winter and restart it at 'the first moon of spring, which corresponded to the calends of March and marked the first day of the year'.[1]

*Statue of Julius Caesar the Calendrical Emperor
(in Rome, Italy) whose date with destiny fell on the
Ides of March of a calendar of his own making.*

Roman farming calendar (first century). Every side shows three months with the star signs, the number of days, hours of the day and night, as well as the agricultural tasks to be undertaken. (Height 65 cm, width 41 cm.)

Papal tweaking – Pope Gregory XIII (Ugo Buoncampagni) 1502–1585, presiding over the commission for reform of the Julian calendar, 1578 (artist unknown).

Over the centuries, piecemeal reforms had been attempted, with the addition of extra months (January and February), but this was only approximately synchronized with the solar year, so an extra month was added every two years, and, in such years, February was shortened. This at best inaccurate system was often modified and manipulated for factional benefit as the pontifices could alter the calendar and so lengthen or shorten the term of officials. By the time of Julius Caesar, 'the civil calendar was out of phase with the seasons by about three months'.[2]

As an empire builder, Caesar needed a tool with which to regulate a growing and increasingly civilized but diffuse and polyglot territory.

He tasked the leading mathematicians, philosophers and astronomers with finding a solution. The result was a solely solar calendar and, in preparation for its introduction, the preceding year was equipped with all the necessary missing days to create a 445-day-year known as the year of confusion. The Julian year of 365 and a quarter days were divided up into twelve months that had nothing to do with the moon, but which would enable equinoxes and the beginnings and ends of seasons to be fixed. Thus, naturally occurring events were integrated into a civil calendar of twelve months, each with either thirty or thirty-one days. The exception was a 29-day February that lengthened to thirty days every third year to compensate for the untidy quarter day. Caesar's successor Augustus subsequently corrected this error and renamed Quintilis (the fifth month) and Sextilis (the sixth) Julius and Augustus respectively to honour the two imperial calendrical reformers.

Although there would be further reforms and corrections in centuries to come, to stabilize calendrical drift, the Julian Calendar is still at the root of how much of the world organises its year today.

As it happened, Caesar implemented his reform just in time: even though he had changed the date, he was unable to avoid the grisly fate that awaited him on the Ides of March in the second year of his new calendar; a fate which a couple of thousand years later would inspire the writers of *Carry on Cleo* to pen the deathless line 'Infamy! Infamy! They've all got it in for me!' for Kenneth Williams' Caesar.

---

1. Dominique Fléchon, *The Mastery of Time: A History of Timekeeping, from the Sundial to the Wristwatch: Discoveries, Inventions, and Advances in Master Watchmaking* (Paris: Flammarion, 2011), p. 70
2. *Ibid.*

The Twilight *of* Antiquity
The Great Clock of Gaza

'*All the known sophists of Gaza in the fifth and sixth centuries are Christian, though their rhetorical works might well have been written by pagans. Two of them, however, are men of earnestness and conviction; we owe to them commentaries on the Holy Scripture and apologetic works. The most brilliant of these Gazan sophists is Procopius.*'[1]

Patrology: The Lives and Works of the Fathers of the Church, Otto Bardenhewer's 1908 anthology of early Christian biographies, runs to many hundreds of densely printed pages. It may not be the liveliest of reads, but it offers insights into the world of late antiquity and, in its short passage about Procopius, vouchsafes us a glimpse of one of the final flickerings of the guttering candle of Hellenistic philosophy that had once illuminated much of the Mediterranean.

To view Palestine-born Procopius as merely a man 'of earnestness and conviction' is to do an injustice to a great chronicler of late antiquity. As legal secretary to Belisarius, the pre-eminent general during the reign of Byzantine Emperor Justinian, Procopius had first-hand experience of military campaigns against Sasanian Persia, the Vandals in North Africa and the Ostrogoths in Italy, where he participated in the siege of Rome. He also lived through the bloody Nika riots against Emperor Justinian as well as an epidemic of bubonic plague. After leaving Belisarius, he moved to Constantinople to write his *History of the Wars of Justinian*, where he appears to have enjoyed some favour at court. It was a court of which he later grew disillusioned, however, writing the *Secret History* – a scandalous exposé that portrays the emperor as demonic and the Empress Theodora as a sexually voracious wanton, still behaving like the prostitute she had once been.

Procopius is also celebrated for his description of Justinian's monumental church, the Hagia Sophia, and a quite remarkable clock that was built in Gaza as it embarked on its brief period of prosperity and intellectual prominence during the sixth century.

Although lost to posterity, the Great Clock of Gaza was one of the wonders of late antiquity. It showed that while Ostrogoths and Vandals may have trampled over

Opposite: A drawing of how the Great Clock of Gaza would have appeared, based upon the the famous Ekphrasis of Procopius, from Über die von Prokop beschriebene Kunstuhr von Gaza: mit einem Anhang enthaltend Text und Übersetzung der Ekphrasis horologiou des Prokopios von Gaza *by Hermann Diels (1848–1922), and published in Berlin in 1917 by Verlag der Königl.*

the civilizations of the past, the arc of the Eastern Mediterranean, over which the long-lived Byzantine Emperor Justinian ruled, remained a place of culture and sophistication capable of creating exquisite instruments of pleasure and elaborate works that united mechanical ingenuity with aesthetic refinement.

Standing about fifteen feet tall,[2] the Great Clock of Gaza dominated the market square in which it was located. Procopius clearly found it a compelling and intriguing device; some say that it was he who inaugurated the public clock. Certainly he described it in such detail that an early twentieth-century scholar was able to make a drawing of it. It must have been a splendid sight to behold, but a

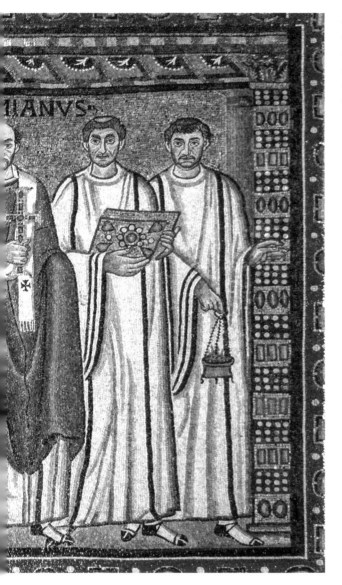

static two-dimensional drawing falls short of capturing the device's impact.

'The clock, erected in a large square, set in motion an elaborate mechanism of mechanical figures. The eyes in the head of the Medusa on the tympanum of the facade moved every hour. Below the Medusa head was a series of twelve doors, in front of which the figure of the sun god was moving about. Every hour a door opened, and out came Hercules with the attributes of one of his twelve mythical labors.'[3]

A mechanical tour de force, there was even 'an eagle with beating wings that wreathed the clock' and 'at night a light moved behind the thresholds of the doors'.[4]

It was a bravura display of the technical capabilities of the age, and the region. Western Europe would not produce a device that neared the Great Clock of Gaza in terms of complexity or ingenuity for many centuries. It represented one of the high points of the tradition of automata, at which Alexandria and Constantinople, the two cities bounding the Eastern Mediterranean littoral, excelled.

In his poem 'Sailing to Byzantium', W. B. Yeats provides a seductive snapshot of the leisured decadence of the world for which automata were made. The poet describes:

… such a form as Grecian goldsmiths make

Of hammered gold and gold enamelling

To keep a drowsy Emperor awake;

Or set upon a golden bough to sing

To lords and ladies of Byzantium

Most famous among the automata experts of antiquity was Hero of Alexandria, scientist and inventor of the syringe, who used his knowledge of water pressure to create amusing automata for the diversion of the leisure class. He could make a miniature tree with a singing bird, an animal automaton that drank from a bowl of water or a satyr that poured water from a wineskin.

The Great Clock of Gaza showed just how far the science had come, as it was an altogether more ambitious object. For example, the base of the clock featured another figure of Hercules that brought his club down on a bronze lion pelt, thus sounding a chime, which, according to horological historian Dominique Fléchon, makes 'the Gaza clock the first known mechanism that chimed the hours'.[5]

And it was not just the clock that was advanced, but the society it served. The mobile ornaments that diverted the 'lords and ladies of Byzantium' were one thing, but this was another entirely. It betokened a bustling, prosperous and settled people who possessed the skills to make such a device and lived in a society that benefitted from the synchronizing effect of a public timepiece. Erecting such a beautiful and inventive object at the centre of commercial life also demonstrated the value that this society placed on time.

Its advanced mechanism aside, the Great Clock of Gaza was pregnant with cultural importance and metaphorical significance.

Byzantium was a Christian empire; it was well known that an angel had appeared to the emperor with a vision of the Hagia Sophia, which he obligingly built, and after a long, increasingly devout life, Justinian would be made a saint. And yet the massive clock in the centre of Gaza bristled with pagan, rather than Christian, symbols.

As admitted with a sorry shake of the head by Otto Bardenhewer, paganism was a stubborn aspect of life in Byzantium, especially as the armies of the emperor reconquered territory around the Mediterranean. Even in the sixth century, the Ancient Egyptian god Amun was worshipped in Libya.

Thus, as well as a dazzling display of technical ability, as Nicole Belayche explains in her essay on the survival of pagan festivals in Gaza, the Great Clock should also be understood as a vehicle for smuggling elements of the old religions into an officially Christian land. It 'perpetuated Hellenism, which Christianity had adopted by separating it from its religious component'. She draws attention to 'Helios in his chariot, with twelve eagles carrying a crown',[6] the twelve labours of Hercules, one for each hour, and so on.

No matter how hard Justinian came down on paganism, it popped up everywhere – even in interior design, where elaborate mosaic floors celebrated the gods of the past. 'In appearance, the towns, henceforth officially Christian, preserved elements of Hellenism, if only in the re-employment of architectonic and decorative pieces.'[7] The Great Clock clearly belongs in this tradition, keeping the memory of paganism alive in Christian times.

But even as Christianity was stamping out the embers of old religious practices, a new monotheism was about to set much of the Mediterranean littoral ablaze. Justinian died at the age of eighty-three in 565 CE. Around five years later, a child was born in the Arabian city of Mecca whose followers would carry his faith around the world with such vigour that within 100 years of his death they would lay siege to Constantinople, overrun the old Roman colonies and come within just a few days' ride of Paris. His name was Muhammad.

'Islam progressed as far in one century as Christianity in seven,'[8] writes Norman Davies in his *History of Europe*. And in adding another striated layer to the cultural geology of Europe, they would assimilate and improve scientific developments they encountered in the territories they overran. Paradoxically, it was by overrunning the old civilizations that the Islamic world preserved the wisdom of the ancient world. These were not barbarians, but men who were as hungry for learning as they were for land, and, as we shall see, they certainly learned from the Great Clock of Gaza.

---

1. Otto Bardenhewer and Thomas J Shahan (trans), *Patrology: The Lives and Works of the Fathers of the Church* (Freiburg im Breisgau, St. Louis, Mo.: B. Herder, 1908), pp. 541–2
2. *Critical Inquiry*, Autumn 1977, p. 101
3. Gerhard Dohrn-van Rossum and Thomas Dunlap (trans), *History of the Hour: Clocks and Modern Temporal Orders* (Chicago, IL.: University of Chicago Press, 1996)
4. *Ibid.*
5. Dominique Fléchon, *The Mastery of Time: A History of Timekeeping, from the Sundial to the Wristwatch: Discoveries, Inventions, and Advances in Master Watchmaking* (Paris: Flammarion, 2011), p. 86
6. Brouria Bitton-Ashkelony and Aryeh Kofsky Koninklijke (eds), *Christian Gaza in Late Antiquity* (Leiden, The Netherlands: Brill NV, 2004), p. 20
7. *Ibid.*, 19
8. Norman Davies, *Europe: A History* (Bodley Head, 2014), p. 253

The Wonders *of the* East
Charlemagne and the Caliph's Clock

On 3 March 2013, Islamist rebels surged into the city of Raqqa. By 6 March, the final units of pro-government forces were overrun, and the northern Syrian city became the first provincial capital to fall to those opposed to the Assad regime. But within a few months the dream of a 'free' Syria had become the nightmare of ISIS, and the name of Raqqa became known around the world as the capital of the newly named caliphate.

It was not the first time Raqqa had served as a caliph's capital: 1200 years ago, Raqqa was where one would have found the court of Caliph Harun al-Rashid, the Abbasid ruler made famous by the tales of *One Thousand and One Nights*, whose 'dominion', says Gibbon in his *Decline and Fall of the Roman Empire*, 'stretched from Africa to India'.[1] For a few years, this city on the Euphrates enjoyed a brief but intense cultural flowering, as he moved his government and court from Baghdad to Raqqa. In the eighth and ninth centuries, this was the most civilized place in the world.

The cultured caliph was the head of a recently established ruling dynasty: the Abbasids had overthrown the Umayyads in the middle of the eighth century, but the Umayyads still ruled the independent Emirate of Cordoba and until 759 were in control of cities in what today is France.

The sophistication of the clock he gave Charlemagne mirrored the sophistication of the court of Caliph Harun al-Rashid, the Abbasid ruler made famous by the tales of The Book of One Thousand and One Nights. *His 'dominion', says Gibbon in his* Decline and Fall of the Roman Empire, *'stretched from Africa to India.'*

Just as the Abbasids were establishing a new order in the East, so a new ruling dynasty, the Carolingian, was emerging in Europe. Acting on the basis that 'the enemy of my enemy is my friend', the newly established Abbasid and Carolingian dynasties established an alliance that became the fulcrum on which European power politics turned, balancing the powers of Byzantium, Umayyad Spain, the Abbasids and the rapidly expanding Carolingian Empire. Harun al-Rashid's counterpart was Charlemagne: they were two of the most remarkable figures of the period between the collapse of the Roman Empire and the beginning of the Middle Ages.

To call these the Dark Ages is not entirely correct, however, as from time to time the court of the Franks in Aachen was illuminated by the splendours of the East. As tokens of his high regard, envoys brought fabulous objects from the 'King of Persia', as the Franks called al-Rashid.

Among the diplomatic gifts they brought were a tent, an elephant, the keys of the Holy Sepulchre and a water clock described in great detail in the Royal Frankish Annals that bore a striking – forgive the pun – resemblance to the Great Clock of Gaza.

The exoticism of the meeting of European and Middle Eastern cultures at the court of the mighty Charlemagne provided fecund ground in which the imaginations of future generations of artists flourished: the moment of the presentation of the famous timepiece particularly appealed to the baroque eye of Jacob Jordaens, who painted 'The envoys of the Caliph Harun al-Rashid offering a clock to Charlemagne' in 1660.

Splendid even in decay – the Baghdad gate at
Raqqa, Syria, captured by the indefatigable
Gertrude Bell in 1909.

It was a 'brass clock, a marvelous mechanical contraption, in which the course of the twelve hours moved according to a water clock, with as many brazen little balls, which fall down on the hour and through their fall made a cymbal ring underneath. On this clock there were also twelve horsemen who at the end of each hour stepped out of twelve windows, closing the previously open windows by their movements.'[2]

The splendour of the clock suggests that giving the time of day was little more than a pretext for the display of mechanical effect and visual delight that the mechanism served up. This sort of thing certainly made a lasting impression on the rulers of the lands that had comprised the old Western Roman Empire.

Such was the novelty of this contraption that it was viewed as a species of magic by the Franks. But in the East, such creations were well known. While barbarian Europe had lost the technology of the Ancients, the great engineering texts of antiquity had been translated from Greek into Arabic and their principles further developed. The water clock had come a long way from Karnak, and it would go further still as the Islamic world continued to work on ever more extravagant clepsydrae for another five centuries.

Gradually this technology found its way to the West, at first through high-level diplomatic missions, and then, as the age of the Crusades dawned, with those who returned from the Middle East with tales of the spectacular feats of engineering they had witnessed there.

Water clocks soon found their way into European monasteries, where, equipped with rudimentary chimes, they tolled out the times of prime, terce, vespers and the other prayers that punctuated the Christian day.

---

1. Edward Gibbon, *The History of the Decline and Fall of the Roman Empire* (London, Strahan & Cadell), p. 94
2. B. Walter Scholz, *Carolingian Chronicles: Royal Frankish Annals and Nithard's Histories* (Ann Arbor: University of Michigan Press, 1970), p. 87

The Missing Link
Su Song's Celestial Clock

*E*leventh-century Europe was not a particularly pleasant place in which to live. Vikings were still raiding and invading the British Isles, as were the Normans, who were also enforcing their rule around the Mediterranean, while in the Iberian peninsula the two great monotheist religions of Islam and Christianity were fighting a war that would last until the time of Columbus.

These were the days when the sword was far mightier than the pen, yet here and there learning was cautiously advancing and beginning to creep beyond the walls of abbeys and monasteries. Although no one can pinpoint exactly when, 1088 is believed to be the date at which free teaching, independent from church schools, was instituted, in Bologna, Italy. It is a date that allows the Alma Mater Studiorum Università di Bologna to claim the title of Europe's oldest university, which 'began to evolve in Bologna in the late 11th century, when masters of grammar, rhetoric and logic began to devote themselves to law'.[1]

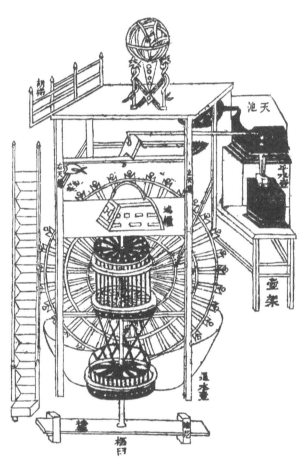

A 'New Design for a Mechanized Armillary Sphere and Celestial Globe' by Su Song, a Chinese astronomer, poet, diplomat and all-round polymath who built his emperor a complex armillary clock housed in a ten-metre-high tower.

The city is understandably proud of its status as the ground zero of the educational movement that would prove crucial to dragging Europe into the Enlightenment, but it would be fair to say that learning, science – and, well, civilisation – was a little more advanced on the other side of the world.

Su Song built a full-scale working model in wood which was tested before the metal parts were cast. Relatively recently 'rediscovered' by Western scholars, Su Song's clock is a 'missing link' between earlier water clocks, in which timekeeping depended upon measuring changes in the water level, and the all-mechanical clock developed in Europe towards the end of the thirteenth century.

A replica of Su Song's Clock at the Gishodo Suwako Watch and Clock Museum.

The year in which the inchoate signs of European university education could be detected in Bologna was also a significant one for Su Song, the minister of justice at the Imperial Court in the city of Bianjing (modern-day Kaifeng).

This was the time of the Song dynasty, under whose rule China enjoyed an explosion of intellectual, scientific and social development. It seems inconceivable that medieval Europe and Song-dynasty China were on the same planet, let alone populated by the same species. For example, those who administered the Song empire needed to pass civil service examinations, a practice not instituted by Great Britain until 1870. The invention of moveable type in the ninth century permitted the distribution of knowledge and the use of paper money; whereas, in contrast, Gutenberg only introduced the printing press to Europe in the fifteenth century, hundreds of years later.

Under the Song dynasty, intellect was prized. Elite status came with education, and even by the standards of the time in which he lived, and the empire that he served, Su Song was exceptionally well-educated. Of commendably broad interests, had he lived in Europe a couple of centuries later he would surely be known to history as a Renaissance man. As well as fulfilling one of the great offices of state, he is described variously as a botanist, zoologist, engineer, architect, antiquarian, cartographer, pharmacologist, doctor, mineralogist, diplomat and poet.

As a young man he had also come top in the astronomy paper of his provincial government exams, and it was in his capacity as an horologist and astronomer that in 1088 he approached the emperor with a wooden model of a pagoda tower. It probably looked like a doll's house, but a serious man like Su Song would hardly have been presenting the emperor with a child's toy.

It was in fact a preparatory model for a large clock and planetary observatory, complete with an armillary sphere and a clock with jacquemarts (automated figures) to strike the time. The model was just the beginning of a feasibility study: the full-size structure would need considerable investment and much effort. It would also have to be built to minute tolerances.

A working model of the water-balance escapement of Su Song's clock. Now recognized as the 'missing link' between the purely hydraulic clepsydra and the all-mechanical escapement clock.

Su Song was a renowned Han Chinese polymath described as a scientist, mathematician, statesman, astronomer, cartographer, horologist, medical doctor, pharmacologist, mineralogist, zoologist, botanist, mechanical and architectural engineer, poet antiquarian and ambassador of the Song Dynasty (960–1279). Su Song was responsible for a landmark development in time-telling.

It was to be powered by running water, but this was no clepsydra; it was instead a *mechanical* clock that used water as its source of energy. Like a paddle steamer or a watermill, a constant flow of running water would pour into a series of scoops or cups on a wheel. Each would take the same amount of time to fill before their weight moved the wheel, thus providing regular motion.

The idea had come to Su Song after a spell as ambassador to the court of the Khitan Liao, a rival state to the north. He had visited the palace to offer good wishes to the Khitan emperor, whose birthday fell on the winter solstice. Su Song wanted to visit on the day his own country's calendar identified as the winter solstice, but the Khitan calendar calculated that the solstice fell a day earlier. The potential embarrassment that this might cause can be easily imagined, but Su Song retained his composure and, as a subsequent chronicler would record, he 'calmly engaged in wide-ranging discussions on calendrical science, quoting many ancient sources of authority, which puzzled the barbarians, who all listened with surprise and appreciation'.[2] Finally, Su Song was permitted to offer congratulations on the day desired (that is, based on the Northern Song calendar). On his return, Su Song reported his experiences to Emperor Shenzong. While pleased that 'his' calendrical assessment of the Emperor's birthday had been successful, he also asked Su which calendar was right. When Su admitted that, in truth, the 'barbarians' had been better at composing the calendar, the Emperor had the erring officials punished.

Su Song had pulled off a major diplomatic coup: imposing the Song calendar on a rival head of state, even though, as he later confided to his emperor, he knew it to be incorrect. His pragmatic cynicism would have impressed Machiavelli, but there was more to this than the preservation of imperial face; the admission of such a calendrical error to a rival power could seriously weaken the state, as Dagmar Schaefer, a historian of pre-Modern Asia, explains:

> *Dynastic houses in the historical region of what is now China ruled according to a heavenly mandate (tianming), which was manifest in the dynasty's ability to foresee and interpret usual and unusual events in heaven, on Earth, and in the human world, to identify units of time, and order the world accordingly. The observation of planetary movements, the movement of the stars, and weather phenomena were all, therefore, integral to court life and central state bureaucracy. Conversely, court and state bureaucracy had an impact on the clock's design.*[3]

An accurate astronomical clock was more than a piece of technological apparatus used to expand human understanding of man's place in the cosmos; it was an instrument of rule, a direct link or hotline to divinity, a conduit through which heavenly wisdom flowed into the Imperial court. Clearly the emperor could not trail behind 'barbarians' in an area of learning at which the Song dynasty had long excelled.

So, Su Song envisaged a multipurpose machine that mimicked the movements of heavenly bodies, gave the times of sunrise and sunset, and indicated the passage of time with a series of figures – a sort of highly elaborate Imperial cuckoo clock.

Construction of Su Song's tower clock coincided with an era of peace and prosperity. The water-powered clock and armillary sphere was a grand project destined to increase the heavenly knowledge of the ruling dynasty and the wider prestige of the empire. A meeting of government chancellors was convened to examine the model and debate the feasibility of such a project.

> *Once they were convinced that the model functioned, court painters began to produce plans for constructing the tower and the armillary sphere. Four months later, the chancellors were asked to 'assemble at the Chonghe Hall to inspect the construction plans of the armillary sphere'. The armillary sphere was then cast in bronze over the next two months, and construction of the tower began.*[4]

Gifted though Su Song was, the plans for the tower clock were not without precedent. Some scholars credit the Han-era engineer and mathematician Zhang Heng with inventing the mechanical, water-driven astronomical clock as far back as the second century CE. Rather more recent was a Song-dynasty precursor to the tower clock: in 977, Zhang Sixun completed 'The Astronomical Clock of Supreme Peace'.

However, Su Song's clock was to be larger, more advanced, and would come to represent the pinnacle of astronomical and mechanical development of the time. There is a kind of poetry in the way that Su Song explained the concept to the Emperor Zhezong:

> *The heavens move without ceasing and so also does water flow [and fall]. Thus if the water is made to pour with perfect evenness then the rotary movements [of the heavens and the machine] will show no discrepancy or contradiction; for the unresting follows the unceasing.*[5]

The 'perfect evenness' was achieved with a system of interconnected tanks, the last of which was always kept at the same height to ensure that water flowed at a constant rate from the spout at its bottom into the scoops of the primary gear wheel. These turned the drive shaft, which in turn rotated the main transmission shaft that ran vertically through much of the building, controlling the movement of the eight gear wheels of the clock.

> *The jack-work wheels performed a variety of functions, their figures either appearing with placards on which the time was marked, or ringing bells, striking gongs or beating drums as they made their appearances in clothes of different colours at the pagoda doorways.*[6]

It must have seemed miraculous. Further poetry and mysticism were imparted by the names given to various components: the pond of heaven (one of the water tanks); the pillar of heaven (transmission shaft); the wheel of heaven (the gear wheel connected to the sphere) and so on.

Emperor Shenzong (Zhao Xu), sixth ruler of the (Northern) Song Dynasty.

'Celestial ladder' was the name given to the famous drive chain, now regarded as the most significant innovation of Su Song's clock. It was responsible for moving the armillary sphere. Su Song described it thus:

> *An iron chain with its links joined together to form an endless circuit hangs down from the upper chain-wheel which is concealed by the tortoise-and-cloud (column supporting the armillary sphere centrally) and passes also round the lower chain-wheel which is mounted on the main driving-shaft. Whenever one link moves, it moves forward one tooth of the diurnal motion gear-ring and rotates the Component of the Three Arrangers of Time, thus following the motion of the heavens.*[7]

Alas, this spectacular machine could not protect the northern part of the Song empire from Jurchen (Tartar) invaders, who in 1125 dismantled the Su Song masterpiece and took it to Beijing. They were unable to re-erect it, however, even with the detailed account left by its creator. Without the physical manifestation of his genius, the wondrous astronomical clock and its maker faded from memory. Moreover, as Western Europe became culturally dominant, the history of horology came to be written with Europe at its centre.

Su Song and his work sank beneath the tide of history and would not resurface again until 1956, when three Cambridge scholars, Needham, Price and Wang Ling, located and translated Su Song's text – thus discovering what one newspaper report aptly described as 'the missing link' in the history of clockwork.

---

1 'The University from the 12th to the 20th century', Alma Mater Studiorum Università di Bologna (https://www.unibo.it/en/university/who-we-are/our-history/university-from-12th-to-20th-century)

2 Quoted in Heping Liu, '"The Water-Mill" and Northern Song Imperial Patronage of Art, Commerce, and Science', *The Art Bulletin*, Vol. 84, No. 4 (December 2002), pp. 566–95

3 Iwan Rhys Morus (ed.), *The Oxford Illustrated History of Science* (Oxford: Oxford University Press, 2017), p. 108

4 *Ibid.*, p. 121

5 Quoted in Heping Liu, '"The Water-Mill" and Northern Song Imperial Patronage of Art, Commerce, and Science', *The Art Bulletin*, Vol. 84, No. 4 (December 2002), pp. 566–95

6 Joseph Needham and Wang Ling, *Science and Civilisation in China, Volume 4: Physics and Physical Technology, Part II: Mechanical Engineering* (Cambridge: Cambridge University Press, 1965), p. 455

7 *Ibid.*, p. 457

The Elephant *in the* Shopping Mall
The Elephant Clock of Al-Jazari

It would be interesting to know what historians and archaeologists of the future will make of the shopping malls of Dubai. Only in the penultimate decade of the twentieth century did the construction of hotels, resorts and shopping centres begin in earnest, transforming a centuries-old trading port into the Hong Kong or Singapore of the Middle East.

Just a generation later, this former stretch of desert has become a land of superlatives: to be home to the world's tallest building and the planet's largest Rolex shop are just two of this glittering nation's proud boasts.

Its shopping centres, however, are perhaps the most remarkable feature of Dubai. Some of these air-conditioned masterpieces of engineering and entertainment are so vast that electric buggies are required to ship exhausted shoppers from the huge Hermes to the palatial Prada. If shopping is the chief leisure activity of the developed world, then Dubai is one of its chief leisure centres.

Dubai knows that man cannot live by shopping alone, so there are also restaurants, food courts and spectacles within these temples of commerce. Dubai is a country that leads humanity with the imaginative diversions it adds to the retail experience. Forget mere cinemas, bowling alleys and gymnasia; the Emirates Mall, for instance, is distinguished by the presence of a huge glass-sided aquarium, where passing shoppers can see colourful fish galore. The same mall also plays host to a ski resort, complete with – what else? – a colony of King Penguins.

In thousands of years, maybe the presence of penguin bones and shards of snowboards will convince archaeologists as yet unborn that climate change was so severe during the early twenty-first century that sub-Antarctic life forms flourished in the Persian Gulf.

When it comes to excavating the Ibn Battuta Mall, these future historians will face another riddle: the presence of a life-size thirteenth-century mechanical elephant surmounted by a towering howdah on which perch humans, dragon-like serpents and birds.

Opposite: The Elephant Clock of Al-Jazari was the apotheosis of the Arabic clepsydra. Conceived with wit, whimsy and healthy splash of eccentricity its complex narrative recalls one of the more involved pages of the illustrated children's book The Cat in the Hat *– one thing balanced precariously on another, balanced precariously on another, and so on.*

A faithful reconstruction of the apotheosis of the Arabic clepsydra, this is the object in which the water clock that began as a calibrated stone bucket in Ancient Egypt reached its apogee. There is a wit and whimsy, as well as more than a touch of eccentricity, about this ingenious creation that makes it irresistible. It is a clock, but it is also the *chef d'oeuvre* of the Islamic world's answer to Leonard da Vinci: Ismail al-Jazari.

This gaudy muddle of cultural appropriation that mixes Medieval Islamic architecture with an Indian elephant, decorative hints from China in the style of serpents, and, of course, Hellenistic/Byzantine water technology, shows just how much of the world Islam had reached by the early thirteenth century.

Al-Jazari was a creative, mathematical and engineering genius; his *Book of Knowledge of Ingenious Mechanical Devices* is a trove of mirthful mechanical contraptions. By the beginning of the thirteenth century, water-powered automata had been around for well over 1000 years, since the days of Philo of Byzantium and later Hero of

Taken from A Book of the Knowledge of Ingenious Mechanical Devices *by al-Jazari, this illustration shows his design for a water clock in the form of an elephant with an Indian driver holding an axe and a hammer – animal welfare was clearly not an issue.*

Alexandria. But the elephant clock demonstrated the heights to which the Arab world had taken the Hellenistic and Byzantine concept of the clepsydra automaton.

Al-Jazari was the most advanced mechanical engineer in a culture that, at the time, led the world in science, medicine and astronomy. The Elephant Clock, which, as the name suggests, represented a pachyderm, was surmounted by an architecturally

Opposite: As this illustration for a 'Figure for Use at Drinking Parties' from his Book of the Knowledge of Ingenious Mechanical Devices *suggests, al-Jazari's ingenuity extended well beyond horology and hydromechanics.*

جنة مقلبه وعلى طرفه
ط وقدار تقع كاس ك
الى ان صارت حافته
بين شفتى النديم
واندفع راسه الى وراه
قليلا بالكاس ثم
يصب الشراب من
طرف مقلب ط الى
قمع ج ويجتمع فى حوض
ت ثم يجف حق ا وفيه
من الشراب بقية
يخرج من المقلب فيقع

وتنحط اليد والكاس وعند مفارقة جانبة الكاس شفة النديم
يعود راسه الى قدامه بسرعة ويخد به مرارا ويستقر وا تمايده
اليسرى وعليها س فانها تنحط كلما ارتفع الشراب فى حوض ص
ورفع العوامة وعليها ع والخيط المتصل بها وهو ملوى على بكرة
تحت كتف النديم وعليها ه ثم يلوى تحت بكرة اخرى
وعليها م وهى تحت طرف مرفق اليد ويرتفع الى فوق ويتصل
شقب ص والنيلوفرة تنحط باليد وعليها ش جنى يك ا اسفل
ساقها باس خذ النديم ولا يزال كذلك حتى يرتفع الشراب الى
حوض ب الى ان يقارب جنية مقلب الحوض وعليه ن يحينيذ

Al Jazari's masterpiece reborn: the elephant clock in the 'Indian' Area of Dubai's Ibn Buttata Mall.

elaborate howdah, housing various representations of men (including a scribe and a *mahout*, or elephant rider), birds and snakes, is the sum and the distillation of all his skill, imagination, intelligence and knowledge into one artefact, which he described in considerable detail for his readers with a passage that recalls one of the more involved pages of the illustrated children's book *The Cat in the Hat* – one thing balanced precariously on another balanced precariously on another, and so on.

Having brought this disparate collection of men, animals and objects together, al-Jazari then brought them to life to tell the time. The scribe's pen moved regularly and in effect acted as the clock's hand. After half an hour, the bird at the top of the howdah whistled and a human figure released the beak of a falcon, which dropped a ball into the mouth of a serpent, the weight of which acted upon the snake like a child getting on a see-saw, bringing the head of the snake near a vase, into which it dropped the ball and returned to its upright position. The *mahout* hits the elephant on the head with an axe in his right hand and raises his other arm with a hammer in his hand. The ball exits the elephant's body, strikes a cymbal and is collected in a tray. The scribe returns his pen to the start position and then, after another half an hour, the bird handler releases the beak of the other bird, the sequence of events is repeated, and a second ball drops into the tray, showing that now an hour has passed.

And the motive force for this complicated three-dimensional maze of interaction? A tank of water concealed beneath the howdah in the elephant's back, in which were arranged a couple of cords and a perforated float. The perforation permitted the float to sink slowly for half an hour, and one of the strings – connected via a system of pulleys and wheels – moves the pen-wielding scribe. When submerged to its lip, the float tilts and sinks rapidly, triggering the theatrical sequence of movements that begins with the release of the ball at the top of the howdah. And it is the movement of the serpent with the ball in its mouth that pulls the cord that brings the float to the surface, ready to commence its next descent. The subsequent movement of the ball once it has left the serpent's jaw activates the *mahout*.

'I have made many varieties of water-clocks using the perforated float at different places and at various times,' al-Jazari wrote, 'and finally have combined them in a single clock, namely, the elephant water-clock.' [1]

It is to be hoped that al-Jazari would be gratified that, eight centuries after its debut, his ultimate achievement was continuing to amaze visitors to the Arabic world… and in particular its shopping malls.

But, at the same time as al-Jazari was bringing the art and science of the clepsydra to a new level of refinement, a completely new technology was waiting, or, perhaps more accurately, 'weighting', to make its debut in the wild and barbarous West.

---

1. Ibn al-Razzaz al-Jazari and Donald R. Hill (trans), *The Book of Knowledge of Ingenious Mechanical Devices* (Dordrecht-Holland/Boston, USA: D. Reidel Publishing Company, 1974), p. 59

A Mechanical
Stairway *to* Heaven
Richard of Wallingford's Astronomical Clock

*ike the first grey glimmerings of the dawn of the Renaissance, the mechanical clock
appears on the periphery of the European experience in the last quarter of the thirteenth
century. By the beginning of the 1270s, a major breakthrough was tantalisingly close,
as Robertus Anglicus, an astronomer from England, recorded in 1271: 'clockmakers
are trying to make a wheel that will have a movement exactly matching the time of one
orbit of the Sun around the Earth in one day, but they cannot accomplish it.'*[1]

In other words, they were attempting
to create what was in effect a twenty-
four-hour clock. By the thirteenth
century, this wheel had clearly been
mastered, as clocks began to appear in
cathedrals and priories.

To the modern mind, the word 'clock'
suggests a visual representation of
time – the hours, minutes and so forth
calibrated on dials with hands. In
the late thirteenth century, however,
the minute was still far too small
a fraction of time to capture on a
mechanical object. Hours presented
enough of a challenge. The typical
'horologe' (derived from the Ancient
Greek for 'hour-teller') was an aural
device, little more than an automated
bell system to summon monks to
prayer at set times.

Their basic appearance belied the
importance of the technological
leap they represented, however.
The first mechanical clocks relied on

*Abbot Richard of Wallingford, his features scarred by
leprosy, gesturing proudly to the complicated astronomical
clock that was his life's work (an illustration from the*
Golden Book of St Albans, *1380)*

This fourteenth-century miniature of Richard of Wallingford, Abbot of St. Albans, shows him hard at work –
his crozier is visible but his attention is clearly focused on the pair of compasses.

the verge and foliot that came between the power source (a dropping weight) and
the gear wheels, which were turned by the dropping weight and in turn drove a
visual or aural indication of time. The verge and foliot took the 'raw' energy of the
dropping weight and turned it into a regular, even supply of energy to drive the gear
wheels smoothly and steadily. In performing this task, the verge and foliot made a
metronomic ticking noise new to human ears. In the late thirteenth century, this was
the sound of the future.

Alas, it is not known who first heard the rhythmic *tick tock* of clockwork that is now
so familiar, as history has not recorded the name of the inventor. History, however,
would certainly hear of Richard of Wallingford.

Replica of Richard of Wallingford's clock, at St Albans cathedral.

In 1292, at around the time those first crude blacksmith-like clocks could be heard tolling their religious message over Edward I's England, a boy called Richard was born to the wife of a blacksmith. By the age of ten, he was an orphan.

Adopted by the Prior of Wallingford, who recognised his potential, Richard was dispatched to Oxford and after six years of study became a monk at St Albans. Three years later he was back at the university, where he would remain for nine years, writing extensively on mathematics, trigonometry and astronomy. He invented an astronomical computing device called the Albion, as well as the rectangulus, an astronomical device used to measure the angles between planets. He also produced a comprehensive study on the construction of clocks and astrological instruments.

Looking back on the exquisite treasures that survive from the abbeys and monasteries of medieval Europe, it is easy to regard these religious houses as citadels of scholarship, sheltered gardens in which beauty flourished, islands of learning that offered refuge from the turbid seas of ignorance and barbarism that had engulfed the continent. Doubtless there were religious institutions just like that, but in the early decades of the 1300s, the monastery of St Albans was not one of them.

Even at their most cordial, relations between civic and clerical life in St Albans were strained. The yoke of the clergy did not sit easily around the necks of the townspeople, who felt their rights were eroded by the Church. They showed their discontent by fishing in the abbot's ponds and flouting the numerous restrictions imposed on them, the most onerous of which being the interdiction on milling their own flour and the obligation to use the monastery's hi-tech and highly expensive mill at their own cost – a source of ongoing discontent for much of the fourteenth century. From time to time, tensions between townsfolk and abbey would flare into naked hostility: monks were assaulted and church property damaged.

The clergy were resented because of the privileges they enjoyed and, under Abbot Hugh of Eversdone, relations reached a new nadir. 'The monastery was besieged by the townsmen of St Albans for forty days, and suffered at least two major assaults. During that period no food was allowed in, and attempts were made to set fire to the buildings. Two of the abbot's men were forced to pay large sums of money to be allowed into the town. The king's bailiff was captured and imprisoned.'[2]

These were tumultuous times. In January 1327, King Edward II was deposed in a coup masterminded by his wife Queen Isabella, who ruled with her lover in the name of the teenaged Edward III, until he in turn overthrew his mother. Richard's return to St Albans in September 1327 coincided, almost to the day, with the death of Abbot Hugh. He immediately began to position himself as Hugh's successor, but his pursuit of arcane sciences was viewed with suspicion. 'There are some', wrote a chronicler, 'who say that he predicted by the constellations of the stars that Abbot Hugh would die and that he would himself become abbot.'[3] If he had made such predictions, he would have been gratified by their accuracy.

Although in a febrile state, the priory of St Albans was a potentially rich prize, its primacy ahead of all other priories in the country ratified by the Pope. For the first few years of his abbacy, Richard asserted his authority: seeing off internal opposition by overcoming plots by fellow priests to oust him; and bringing the townspeople into line by taking away their private mills, restoring one of the abbey's chief revenue streams.

The money was badly needed. St Albans may have had the 'longest nave in Christendom',[4] but columns on its south side had collapsed in 1323 and Bishop Hugh had lacked the funds to do anything about it. Having restored the abbey

Astronomical dial, bell and gears of Wallingford's clock in place in St Albans Cathedral.

finances, Richard now had the funds to carry out much-needed repairs. Instead, he chose to use those funds to realise a plan that had doubtless been forming in his agile mind for some years. He may have written, drawn and dreamed while in the libraries of Oxford, but now he was the master of a prominent religious institution, he exercised real power and had access to real resources.

He was in a position to put his theories to the ultimate test and build the most advanced horologe of the age, one that unified the functions of astrological instruments and the newly invented mechanical clock. Of course, it could be presented as evidence of a desire on Richard's part to create something that declared the prestige of St Albans, dedicated to the glory of God, linking the sublunary existence of man with the celestial realm by means of a mechanical machine … Then again, it could just be that, having come so far from the forge where he had been born, the Abbot of St Albans wanted to reach out to the stars, to God himself, with a machine of his making.

His obsession did not go unnoticed, however, and his brother clerics complained to each other about the cost of what some appear to have seen as a vanity project when the church was in need of repair. Even King Edward III questioned the expense when he came to the monastery to pray. But Richard had a ready answer for the royal rebuke, as the chronicler of his abbacy recounts: '[He] replied, with due respect, that enough Abbots would succeed him who would find workmen for the fabric of the monastery, but that there would be no successor, after his death, who could finish the work that had been begun. And, indeed, he spoke the truth, because in that art nothing of the kind remains, nor was anything similar invented in his lifetime.'[5]

However, while the astronomical clock may have been invented during his lifetime, it was not completed in it. Illuminated manuscripts illustrated with images of Richard disclose little of his character, but act instead as a visual *curriculum vitae*. For example, we see a young Richard with a tonsured head, bent over a desk, surrounded by books, crozier in the crook of his left arm, a pair of dividers in the other: a picture of scholarly zeal with the trappings of monastery life. But the depiction of Richard in later life shows a very different man: here he is in his dignity and pomp as premier Abbot of England: mitre on head, crozier in hand, he points with pride to the gleaming face of a tower clock. But his own face is disfigured with sores and his nose is little more than a contusion between his eyes and his mouth, more carbuncular snout than recognisable human facial feature.

Richard was a dead man walking. He had contracted leprosy, most likely on a trip to Avignon to meet the Pope. Increasingly physically handicapped, the end of his life was hastened when lightning struck his chamber, and on his death in 1326 work on the horologe ceased. It was only during the abbacy of Thomas de la Mare between 1349 to 1396 that the horologe was completed and remained in the south transept of the abbey, where, even in the sixteenth century, it was still capable of stunning those who looked upon it. John Leland the antiquarian was awestruck when he saw it in the 1530s. Describing Richard as 'the greatest mathematician of all in his time', Leland wrote, he 'constructed such a fabric of a horologium, with great labour, at greater cost, and with still greater skill and, in my opinion, it has not its equal in all Europe; one may look at the course of the sun and moon or the fixed stars, or again one may regard the rise and fall of the tide, or the lines with their almost infinite variety of figures and indication.'[6]

In truth, Leland was probably baffled by these important-looking geometrical indications, and his head would have swum trying to understand how it showed the positions of the heavenly bodies in real time and predicted when lunar eclipses would take place. There was, moreover, a feature called the Wheel of Fortune (thought by later scholars to have been an automaton that reminded those who saw

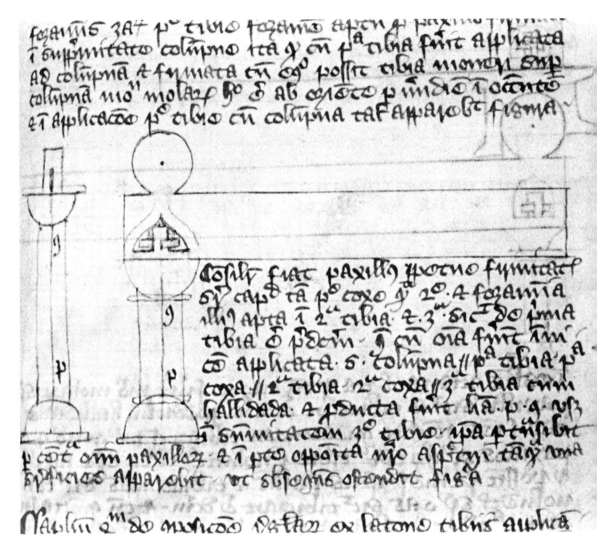

A section of Richard of Wallingford's treatise on the rectangulus, discovered at the Bodleian library (the treatise was edited and translated by J. D. North in 1976).

it of the capriciousness of fate) that only deepened the mystery of an object designed well over a century before the Wars of the Roses, yet using technology that was still, 200 years later, beyond the wit of even educated men to fathom.

But, worse than not understanding it, Tudor England destroyed Richard of Wallingford's horologe during the Reformation, and only in the 1960s did documents come to light enabling replicas to be rebuilt. Such was the conceptual power of the astronomical clock, however, that its reputation survived for four centuries based solely on a handful of accounts written posthumously.

Towards the end of his life, enfeebled, leprous and bedridden in his lightning-struck room, as he prepared to meet his God, according to one eminent scholar of medieval horology, Richard 'is said to have regretted devoting more time to scientific studies than to theology. The loss to theology, however, was a considerable gain to medieval astronomy and technology.'[7]

Had he followed a life devoted to theology and prayer, he may well have faced his end with more equanimity and less fear for his immortal soul. But, as it is, history has bestowed upon him another sort of immortality – preserving him from the obscurity that has long shrouded many more overtly pious clerics.

1. Quoted in Francois Chaille, *The Beauty of Time* (Paris: Flamarrion, 2018), p. 110
2. John North, *God's Clockmaker: Richard of Wallingford and the Invention of Time* (London: Bloomsbury, 2005), p. 124
3. *Ibid.*
4. *Ibid.*, p. 14
5. Quoted in Silvio A. Bedini and Francis R. Maddison, 'Mechanical Universe: The Astrarium of Giovanni de'Dondi', *Transactions of the American Philosophical Society*, Vol. 56, No. 5 (1966), pp. 1–69
6. *Ibid.*
7. Silvio A. Bedini and Francis R. Maddison, 'Mechanical Universe: The Astrarium of Giovanni de'Dondi', *Transactions of the American Philosophical Society*, Vol. 56, No. 5 (1966), pp. 1–69

The Mechanical Cock
that Crowed
The Marvel of Strasbourg

The beak is chipped. There is what looks like a suggestion of woodworm at the throat. The wings have a forlorn air. A once splendid but now time-dulled tail feather has been snapped off. And a spur is missing from one leg.

To mistake this replica rooster for a weathervane that has seen decidedly better days would be to do a criminal disservice to what, in the eyes of the Musée des Arts Décoratifs in Strasbourg, is a medieval mechanical masterpiece. Made from wood and iron by some long-dead and unnamed craftsman in around 1350, it is, says the museum, the oldest surviving example of a European automaton.

A century before the birth of Leonardo Da Vinci, this battered-looking bird was the crowning ornament of the Marvel of Strasbourg: a public clock that was the talk of medieval Europe.

As has been seen in Hellenic Gaza and at the court of Charlemagne, a tradition of embellishing the function of timekeepers with automata had long existed. But, by the mid-fourteenth century, the water clock was yesterday's technology – as anachronistic as an abacus in the age of the electronic calculator.

Visitors admiring the astronomical clock in the Cathedral of Strasbourg, France.

Mechanised clocks were in, and once the technology had been mastered, it would prove crucial to the defining cultural, social and economic phenomenon of medieval Europe: the rise of the city. The great cities of Europe lost no time in erecting large mechanical clocks and in turn came to be defined by them. Physically and figuratively at the centre of urban life, the grand municipal clock began tolling its unifying message of community. When the Church of San Gottardo in Corte in Milan acquired its '*campanile delle ore*' (literally 'bell tower of the hours'), the first public clock in the city, the surrounding streets became known as the '*Contrade delle ore*' ('Quarter of the hours'). In London, the sense of identity bestowed by a public clock would be so powerful that those born within earshot of the bells of St Mary-le-Bow, which still sound every quarter-hour, would call themselves Cockneys.

The sound of the time could be heard by serf and noble alike; it entered every home, workshop, palace, mansion, guildhall, court and counting house. These horological behemoths generated the societal adhesive of synchronization that enabled civic institutions, places of worship, places of commerce and domestic households to order and schedule their activities by the sound of the bells.

As the night of the Dark Ages receded, the weight-driven mechanical clock presaged the dawning Renaissance. With its clock in the St Gottardo Church, Milan got off to an early start in the 1330s, but soon any city state with the slightest sense of self-respect made sure it could boast an impressive tower clock: Modena had one in 1343; Jacopo de'Dondi completed an astronomical clock for the Prince of Carrara in Padua a year later; and Monza followed in 1347.

Ever more ambitious and elaborate clocks combining the time of day with astronomical and astrological displays as well as automata became urban status symbols. In later centuries, cities would burnish their reputations with ever more impressive museums, railway

A century before the birth of Leonardo da Vinci, this battered-looking cockerel was the crowning – and crowing – ornament of the Marvel of Strasbourg: a public clock that was the talk of medieval Europe. Made from wood and iron in around 1350, it is said to be the oldest surviving example of a European automaton.

stations and airports, but in fourteenth-century Europe it was these immense public expressions of the clockmaker's art that signalled a city's wealth, prestige and technological sophistication. So jealously guarded were the secrets of the clock tower that a rumour began circulating that, upon completion of the work, the clockmaker was blinded to prevent him making something as good, or even better, for a rival.

When it came to automata, one of the most popular personifications of horological function was the jacquemart, a human-shaped bell-striker; a sort of early robot. Such was the significance of a lavish timepiece to a city that when Philip II, Duke of Burgundy, sacked Courtrai in Flanders in 1382, he dismantled the city's clock, jacquemart and all, and carted it off to his capital Dijon, where it was installed at the cathedral: a symbol of the emasculation of one city and the glorification of another.

As the city's most important building, the cathedral was a fitting location for such a contraption, but there are cathedrals, and then there are *cathedrals*, and by the middle of the fourteenth century, one of the latter sort was rising above the plains of Alsace.

Even unfinished, the towering Cathedral of Strasbourg overshadowed the city and dominated the surrounding area from the Black Forest in the east to the Vosges Mountains in the west. Once completed, it was the world's tallest building until well into the nineteenth century. It is regarded as one of the masterpieces of Gothic architecture, and its construction would take centuries, occupying generations

A sixteenth-century woodcut of the astronomical clock in the Cathedral of Strasbourg by artist Tobias Stimmer, who was briefed to produce an appropriately all-encompassing decorative scheme for the clock showing 'everything from history and poetry, sacred texts and profane ones, in which there is or can be a description of time'.

of craftsmen and labourers. In sunlight the sandstone almost seemed to glow, and for many years its beauty imprinted itself on the imagination of poets and novelists, inspiring them to ever dizzying heights of hyperbole.

Its effect on medieval travellers toiling over the landscape, drawn mothlike to this towering vision of heaven on Earth, would have been indescribably powerful. In a film, the sight would be accompanied by swelling organ music and the sound of heavenly choirs. Having attained the luminous citadel of Christ, passed beneath its rose window and entered through the richly carved West Front, the ecstasy of the senses would continue as a world of beauty and wonder opened up before the visitors in the cavernous interior.

Built between 1352 and 1354, almost a century before the completion of the cathedral, the Three Kings Clock, with its ecclesiastical calendar and automated astrolabe dial (a mechanical device used to measure the height of stars above the horizon and identify planets to perform navigational calculations) was an integral part of the spectacle. Its carillon mechanism filled the air with religious music and, as its name suggests, its automata comprised nothing less than the mechanical recreation of the Nativity, involving figures representing the three Magi paying homage to the Virgin and child. And yet, even as the wise men handed over the gold, frankincense and myrrh to the infant, the eerie portent of the Son of God's betrayal by his apostle came from the cockerel, which in those far off days flapped its wings, opened its beak, stuck out its tongue and, by means of a reed and bellows, crowed.

In an age when literacy was far from universal, this robotic recreation of the birth of Christ was a daily reminder of the Christian miracle and its central place in the cosmos, as shown on the astrolabe dial. Furthermore, an illustrated panel linked signs of the zodiac with body parts, indicating propitious or dangerous times for that medieval medical panacea: bloodletting. Thus, man's very wellbeing was demonstrated to be inextricably linked to the movements of the celestial bodies that roamed the heavens, and, of course, by implication, the ruler of those heavens.

The clock was central to cathedral life, and it is suggested that Passion plays given in the Cathedral depicting the suffering and death of Jesus 'were coordinated with the striking of the hours of the clock and the performance of its automatons'[1] – most significantly, the poignant crowing of the rooster that coincided with Peter's third denial of Christ.

For a couple of centuries, therefore, visitors to Strasbourg Cathedral were amazed, moved, entertained, instructed and enlightened by the great clock – until the rooster fell silent and the clock into disrepair. By the 1570s, the Marvel of Strasbourg was ready for its second incarnation.

Conrad Dasypodius was one of the foremost mathematicians of the sixteenth century, and in his fecund mind the Strasbourg clock was reborn as the ultimate scientific object, into which was crammed almost the entire canon of mathematical, mechanical and astronomical scholarship of the age.

Between 1571 and 1574, the cathedral's soaring transept echoed to the sound of work as a new and even more… well… marvellous marvel began its rise from the cathedral floor. Isaac and Josias Habrecht were the clockmakers responsible for making Dasypodius's encyclopaedic machine a mechanical reality. Further drama came from the brush of mannerist painter Tobias Stimmer, who was briefed to produce an appropriately all-encompassing decorative scheme showing 'everything from history and poetry, sacred texts and profane ones, in which there is or can be a description of time'.[2]

As such, and even though at the time the cathedral was a place of Protestant worship, its appearance made it something of a precursor to the baroque style of ecclesiastical architecture that emerged in the latter part of the century, following the Council of Trent and the Counter-Reformation. Just as elaborately painted church ceilings presented a vivid, cinematic view of a heaven that tumbled with angels and saints who almost seemed to move against a dramatic *trompe l'oeil* cloudscape, so the ornament-crammed clock of Conrad Dasypodius vouchsafed the cathedral's congregation a glimpse inside the very mind of God.

This towering sixty-foot-high creation – the single word 'clock' is almost misleading in its simplicity – was a technical tour de force that served up an extravagant banquet for the eyes. Visually astonishing, it was unlike anything the eyes of man had seen before. Planets moved, eclipses were predicted, a perpetual calendar indicated leap years, the dates of moveable religious feasts were displayed. Interestingly, and not unlike the Great Clock of Gaza 1000 years before, older religions were invoked too; the gods of antiquity were represented next to the weekday named after them.

In contrast to the eternal nature of the Church, the clock was full of reminders of the swiftness with which human life passed. For instance, the quarter hours were struck by a succession of figures: a child with an apple for the first quarter; an adolescent with an arrow for the second; followed by a man with a staff for the third; and finally an aged figure who strikes the bell with his crutch.

Automata brought to life the ineluctable forces of death and time. The gargantuan timepiece served, says Princeton professor Anthony Grafton, 'as an animated almanac and embodied the all-destroying force of time and change'.[3]

Utterly unencumbered by modesty, Dasypodius contended that it offered a real-time display of the universe, showing 'the century, the periods of the planets, the yearly and monthly revolutions of the sun and moon'.[4] And yet even as the clock

One of the masterpieces of Gothic architecture: in sunlight, the cathedral of Strasbourg almost seemed to glow, and for centuries its beauty imprinted itself on the imagination of poets and novelists.

was completed and revealed to Strasbourg's rapt and dazzled worshippers, scientific understanding of the cosmos was moving in a direction that would discredit the received wisdom that placed the Earth at the centre of the universe.

The second coming of the Marvel of Strasbourg was one of the last great technical achievements of the geocentric age, and although it celebrated the Ptolemaic vision of the planetary system, the great clock also shows evidence that Conrad Dasypodius was hedging his bets – giving a cautious welcome to the new heliocentric theory by featuring an image of Copernicus amongst the clock's profuse decoration.

Visitors to the Cathedral could be easily forgiven for missing the portrait of the heretical Polish astronomer, however, as there was simply so much else competing for their attention. The clock dwarfed human viewers and seems to have been the size of a small church itself: it comprised a large central tower surmounted by an elaborate structure that was part crown, part mitre, flanked on one side by a helter-skelter-like spiral staircase and on the other a slightly smaller tower topped by the fourteenth-century cockerel, which, restored and returned to a prominent position, was crowing and flapping its wings again.

The clock was world famous, and during the late sixteenth century its renown spread by engravings. Those who could afford more than an engraving commissioned its maker, Isaac Habrecht, to build intricate astronomical carillon clocks inspired by the Marvel of Strasbourg for their own collections.

Portrait of Charles of Lorraine during the siege of Strasbourg, the Cathedral clearly visible in the distance dominates the landscape.

Destined for the *wunderkammers* of status-conscious Renaissance princes, these were considerable objects in their own right, as the 1.4-metre-high example currently housed in the British Museum demonstrates.

Over the centuries that followed, time once again exacted its toll on the Marvel of Strasbourg, and by 1789 the clock was silenced. Given that Europe was convulsed by the French Revolution and then the Napoleonic Wars, the clock's custodians can be forgiven for neglecting to repair Dasypodius's work immediately.

In the 1830s, however, the much-needed restoration work began, and in January 1843 the *Illustrated London News* reported that the 'interest of this master-piece of mechanism (sometimes called the "Monster Clock") has "taken a new turn" by its having been recently repaired, and exhibited to the Scientific Congress, held a short time since at Strasburg.'[5] The Marvel of Strasbourg was back in business, and after nearly five centuries was still of interest to the scientific community.

While not as radical as the rebuilding by Dasypodius, Jean-Baptiste Schwilgué undertook such a major overhaul that it is customary to talk of this as the third great astronomical clock to occupy Strasbourg Cathedral, and it is Schwilgué's 'interpretation' of Dasypodius's clock that greets visitors today.

According to the *Illustrated London News*, Schwilgué's most notable enhancement was a 'mechanism added by which at midnight, on December 31, the moveable feasts and fasts range themselves in the calendar, and the succession they will appear in during the next year'.[6] There was also a Copernican planetarium, an indication of sidereal (astronomical) time and much more besides.

But Schwilgué has taken care to preserve the old favourites too: the eclipse indication, calendar and astrolabe functions remained. The clock's restoration refocused attention on a centuries-old oddity, the complexity of which, allied to its baroque appearance, appealed to the Victorian mind with its love of the curious and bizarre – and above all this meant the automata. The chief attraction in the nineteenth century, as it had been in the fourteenth, was the tableau vivant of the Nativity performed by the marionettes.

The *Illustrated London News* relished the way that 'the days of the week are represented by different divinities, supposed to preside over the planets, from which they are named. The divinity of the current day appears in a car, rolling over the clouds, and at night gives place to the succeeding one. Before the basement is displayed a globe, borne on the wings of a pelican, round which revolve the sun and moon, the mechanism being in the body of the bird.'[7]

The cathedral clock at Strasbourg has captured the imagination of writers and artists over the centuries.
It even appeared in the Victorian sensation novel, Armadale, *by Wilkie Collins. Here, the Major, an amateur*
horologist, sees his clock inspired on that of Strasbourg malfunction spectacularly.

Watercolour of the astronomical clock, by Massimiliano Pezzolini (born 1972). From a four-part series 'Strasbourg. Die Astronomische Uhr'.

As well as the four ages of man to sound the quarters, the grim reaper also put in an appearance. 'Death, who is expelled by our Saviour rising from the grave, who, however, permits Death to sound the hour,'[8] using a bone as his striker. Angels, too, play their part, one using a sceptre to hit a bell, the other turning an hourglass at the hour's end. A carillon at the top of the clock's main tower played a number of tunes and a new cockerel brought proceedings to close, crowing and flapping its wings.

Those two emblematic Victorians, Charles Dickens and Wilkie Collins, were delighted by what the latter called the 'fantastic puppet show'.[9] The clock made a particularly deep and lasting impression on Collins. In his 1866 sensation novel *Armadale*, the Strasbourg clock and its 'puppet show' provides the basis for one of the book's tragi-comic scenes. Retired soldier and amateur horologist Major Milroy has spent years building a clock that draws upon aspects of the Strasbourg Marvel, but instead of the religious music, the contraption strikes up a military march and the automata enact the changing of the guard, or at least they try to: the device malfunctions resulting in an absurd 'catastrophe of the puppets'.[10]

As well as influencing one of the prototypical sensation novels of the nineteenth century,[11] the attention paid to the new and improved Strasbourg clock even re-ignited the inter-city horological rivalry of the Renaissance, when the citizens of Beauvais, which also possesses an impressive astronomical clock, claimed theirs was superior to that of Strasbourg. (Perhaps Beauvais envied the trail of tourists who came, and still come, to see the Marvel.)

The Alsace city was quick to correct such a grave misapprehension and in 1869 the debate even spilled onto the pages of the *Scientific American*: 'I could not suppress a smile at the catalogue of indications said to be shown by the Beauvais clock, for our cathedral clock shows all these and some besides,'[12] observed Strasbourg's Mr Steckelburger.

According to this splendidly named defender of Strasbourg's honour, the city fathers of Beauvais had made the schoolboy error of comparing their clock to the specifications of the Marvel of Strasbourg Version 2.0, conceived by Dasypodius, rather than the Schwilgué incarnation.

According to Steckelburger, the key to the superiority of the Strasbourg clock was the slowness and infrequency with which some components moved. Beauvais was very proud of a leap century indicator that functioned once every four centuries. However, this was positively fast-moving when compared to what Strasbourg's clock could do…

'The Beauvais clock makes a change in every fourth century; a great merit!' he concedes charitably, only to add, with a paramological flourish, 'But ask an astronomer what is meant by the precession of the equinoxes. He will tell you it is a movement in the stars describing a complete revolution round the earth in the space of about 25,000 to 26,000 years. Well, Sir, in the Strasbourg clock is a sphere following exactly this motion, and whose rotation is of that kind as to ensure one revolution in 25,920 years.'

Moreover, Steckelburger explains reassuringly that the Strasbourg clock did not expect visitors to hang around the cathedral waiting for this component to complete its task: 'The thing can be measured and indicated; it is unnecessary to await its accomplishment: it would be too remote.'[12] But even without being able to see this horological function, there remains much about the great clock to enchant and amaze.

1. F. C. Haber, 'The Cathedral Clock and the Cosmological Clock Metaphor', in J. T. Fraser and N. Lawrence (eds), *The Study of Time II* (Berlin Heidelberg: Springer-Verlag, 1975)
2. C. Dasypodius, *Heron mechanicus; seu de machanicis artibus atque disciplinis. Eiusdem horologii astronomici Argentorati in summo Templo erecti descriptio* (Strasbourg, 1580), quoted by Anthony Grafton, 'Chronology and its Discontents in Renaissance Europe: The Vicissitudes of a Tradition', in Diane Owen Hughes and Thomas R. Trautmann (eds), *Time: Histories and Ethnographies* (Ann Arbor: University of Michigan Press, 1995)
3. Anthony Grafton, 'Chronology and its Discontents in Renaissance Europe: The Vicissitudes of a Tradition', in Diane Owen Hughes and Thomas R. Trautmann (eds), *Time: Histories and Ethnographies* (Ann Arbor: University of Michigan Press, 1995)
4. C. Dasypodius *op. cit.*
5. 'The Great Clock at Strasburg', *Illustrated London News*, 28 January 1843
6. *Ibid.*
7. *Ibid.*
8. *Ibid.*
9. Quoted by Fred Kaplan in *Dickens: A Biography* (New York: William Morrow, 1988), p. 292
10. Wilkie Collins, *Armadale*, 'Book the Second/Chapter VI: Midwinter in Disguise'
11. See Lisa M. Zeitz and Peter Thoms, 'Collins's Use of the Strasbourg Clock in Armadale', *Nineteenth-Century Literature*, Vol. 45, No. 4 (March 1991), pp. 495–503
12. 'The Great Clock of Beauvais Cathedral and the Strasbourg Clock', *Scientific American*, 21 August 1869

The Lost Wonder *of the* Renaissance
De'Dondi's Astrarium

It is one of the great lost artefacts of the Middle Ages. When it comes to mechanics, science and clockwork, the Astrarium of Giovanni de'Dondi occupies the same sort of cultural space as the Amber Room of Catherine the Great. Indeed, just the famous Amber Room was reconstructed during the late twentieth century, so, at more or less the same time, were functioning replicas of the Astrarium. It captures the imagination like a Loch Ness monster – and it actually existed.

The Astrarium leaves a tantalising trail through history, in the form of drawings, sightings, letters, accounts and other collateral materials. It was possessed by princes, even an emperor, and then it disappears, leaving only footprints in the sand of history.

The mechanical clock rivals the later printing press for the title of the defining invention of the Middle Ages. Both were tools that enabled Europe to haul itself out of the gloom of the Dark Ages, to bask salamander-like in the sun of the Renaissance. And in the medieval world, no clock was more evocative and miraculous than the Astrarium of Giovanni de'Dondi.

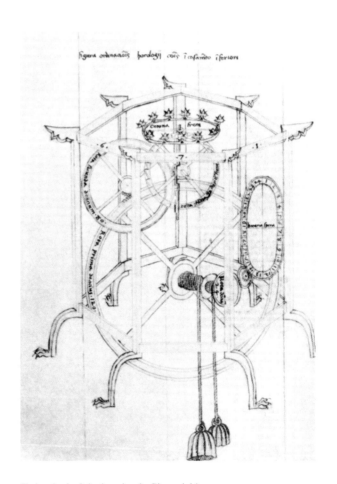

Design sketch of the Astrarium by Giovanni de' Dondi, a fourteenth century professor of astronomy. An astrarium is also known as an astronomical clock and planetarium. One of the great lost artefacts of the Middle Ages, the astrarium of Giovanni de' Dondi was a scientific miracle that was regarded by some as the ultimate scientific achievement of its time, surpassing the works of antiquity and posterity.

When it comes to mechanics, science and clockwork, the Astrarium of Giovanni de'Dondi occupies the same sort of cultural space as the Amber Room of Catherine the Great. As the Amber Room was reconstructed during the late twentieth century, so functioning replicas of the Astrarium were built, such as this one by Thwaites and Reed limited.

Except, to call it a mere clock is to beggar its achievement. The poet Petrarch, that humanist cornerstone of the Renaissance, who saw it at the time, said that only 'uneducated people'[1] could describe it as a clock. Those who were lucky enough to lay their eyes on the Astrarium found it seared itself onto their minds and went away changed by the experience.

'I saw again the globe clock,' wrote one such to its inventor, 'which you made with your hands and which you brought out to take shape from the deepest recesses of your mind: to me it is a magnificent work, a work of divine speculation, a work unattainable by human genius and never produced in generations past; although Cicero tells how Posidonius had constructed a sphere which revolved, showing, through the sun, moon and five planets, what happens in the heavens at night and during the day, I do not believe that there was such competency in art at that time, nor was there such mastery of skill as is shown in this. I do not believe that any of posterity can make it or excel it, since in the passing of time we do not see such sublime growth of genius.'[2]

Heady stuff: never equalled even in classical antiquity, the period revered by the Renaissance, without peer among its contemporaries, and unlikely to be matched in times to come. It was a potent symbol of the genius of man and the development of his understanding of the cosmos and his place in it.

The common concept of the 'clock' in this era had about as much in common with the Astrarium as Bleriot's channel-crossing monoplane has with Apollo 11.

The Astrarium was a metre high, comprising two sections set in a heptagonal frame. The seven panels at the top featured dials that looked like astrolabes and showed the movements of the sun, moon and the then known planets: Mercury, Venus, Mars,

Gian Galeazzo Visconti, first Duke of Milan and noted patron of Dondi. The coloured woodcut is from a guide to the city published by the municipality of Milan in 1906.

Jupiter and Saturn (this, of course, reflected the geocentric, Ptolemaic view of the universe held at the time).

The lower part included a twenty-four-hour clock, dials that indicated dates of fixed and moveable religious feast days, a dial that displayed the intersections of the solar and lunar orbits, and two further displays: one for the time of sunrise and the other for the time of sunset in Padua.

A true wonder of the medieval world, it was the subject of secular pilgrimage by scholars, philosophers, astronomers and leading personalities of the age, all of whom believed there was 'no written or other human record that in this world there ever was made such a subtle or so solemn an instrument of the celestial movements, as the aforesaid clock; the subtle skill of the aforesaid Master John, who with his own hand forged the said clock, all of brass and copper, without assistance from any other person, and did nothing else for sixteen years.'[3]

Master John, or Giovanni de'Dondi, was one of the most remarkable men of the Middle Ages, but to say that he did nothing other than cut toothed wheels, ponder gearing ratios and map the movements of the heavens is to present a picture at odds with what is known of the man. The late eighteenth-century statue in Padua's vast Prato della Valle swathed in robes recalling a toga, the fingertips of his left hand resting delicately on an armillary sphere, suggests a cloistered intellectual. However, it seems he was not the prototypical medieval monastic scholar in the mould of Wallingford, but closer to the colourful figures who so enliven the Renaissance with their ability to turn their talents with equal facility to court life, mathematics, diplomacy, astronomy, politics, medicine and so on, fuelled by their insatiable appetite for knowledge and understanding.

Perhaps what history has come to know as Renaissance men believed that they lived in a time when all knowledge could be concentrated in one mind:

When Francesco Sforza became Francesco I of Milan, establishing a new dynasty to succeed the Visconti family, the Astrarium was still considered one the greatest ducal treasures, even though about a century had passed since its manufacture. (Portrait c. 1460. Found in the Collection of Pinacoteca di Brera, Milan.)

if the body in which it was housed managed to work hard enough and stay alive long enough. This was quite a considerable 'if', given, for example, that excitement around the proliferation of the weight-driven clock across medieval Europe was rather eclipsed by the Black Death in the late 1340s, which relieved the continent of between a third and half of its population – a feat that neither the First nor the Second World War was able to achieve.

Even as a pile of fourteenth-century spare parts and scrap metal, the Astrarium could still fire the imagination. It even caught the eye of Charles V at the time of his coronation as Holy Roman Emperor in Bologna – one of the greatest ceremonial occasions of the sixteenth century. Charles V, Holy Roman Emperor (with his English water dog), 1532, by Jakob Seisenegger (1504/5–1567), Austria, sixteenth century.

Giovanni was not, perhaps, a Leonardo, but it could be argued that he was a prototype, an early model so to speak, of the great Da Vinci. He was born into a family of high achievers. His grandfather had been a physician and his father, Jacopo, born in around 1290, distinguished himself as a lecturer at the school of medicine in Padua, as an author of medical treatises, and with the design and construction of an astronomical tower clock in Padua for the Prince of Carrara.

Fourteenth-century Padua witnessed a remarkable flowering of the arts and sciences under the rule of the Carrara family, and Jacopo's son Giovanni was one of its brightest blooms. By the early 1350s, he was professor of medicine at Padua University and his career took off like a rocket. By the end of the decade, he was a member of the faculties of astronomy, logic and philosophy as well. His reputation as a scholar grew beyond Padua, and during the 1360s he lectured in Florence before being sent as ambassador to Venice in 1371. The following year he was one of five citizens to serve on the boundary committee set up to establish borders with the Venetian Republic. It appears that the commission was not successful, as that year he was among the Paduans who voted for an ultimately futile war against Venice.

He lost the patronage of Francesco da Carrara, but, much in the way that scientists during the Cold War were lured into defecting across the Iron Curtain, Giovanni soon found another protector: Gian Galeazzo Visconti of Pavia, to whom he presented the Astrarium.

Simultaneously brutal and enlightened, Visconti inherited the rule of Pavia from his father, overthrew his uncle the ruler of Milan, and seized Verona, Vicenza, Piacenza and for a while Padua, welding them into a powerbase from which he launched attacks against Bologna and Florence. He accumulated cultural treasures just as avidly as he added new cities to his realm, and de'Dondi's Astrarium was the ultimate trophy.

Unsurprisingly, Visconti 'loaded him with benefits and honors'.[4] Even in his seventies, de'Dondi was still involved in the powerplays of the northern Italian courts, and it was while visiting the Doge of Genoa in 1389 that he fell ill, dying in Milan on 22 June of that year. During a busy life by any standards, he managed to find the time to produce written treatises, marry twice, father nine children and – of course – invent the defining mechanical object of his time.

As well as the knowledge to design it and the skill to assemble it, it was the imaginative power to conceive of such a machine that is most remarkable. As visitors travelled to marvel at the ingenuity and exigence of the object, they were also paying homage to the scope of a mind that could come up with such a device – the aim of which was, said de'Dondi, to bring 'common appreciation to the noble study of astronomy, which had been troubled and weakened by astrological fallacies, rendering many of the early studies on the movements of planets absurd'.[5]

Visconti placed the machine at the centre of the celebrated frescoed ducal library, where light flooded in through high windows and glinted on the silver chains securing the leather- and velvet-bound volumes. Petrarch studied here, as, later, did Da Vinci, who made detailed sketches that some believe to be of the Astrarium's planetary dials. Indeed, the Astrarium and the library became bound together in the cultural imagination of the time.

When the Sforza dynasty assumed control of Milan, the Astrarium loomed large in the concerns of the Sforza duke, who dispatched one of his chancellors to take an inventory of the library of Pavia and gave him instructions to find someone to restore the clock now that its creator was dead.

Even if working imperfectly, in 1460, a century or so after it was made, the Astrarium was still impressive enough for mathematician Regiomontanus to record its power of drawing important visitors to the castle in Pavia: 'In order to see it, innumerable prelates and princes have flocked to that place as if they were about to see a miracle, and indeed not without cause, so great, and indeed so unusual, is the beauty and utility of this work, that there is no one who does not admire it.' [6]

By the 1520s it had fallen into disrepair – if not to pieces, then certainly into a state of incompleteness. Yet, even as a pile of fourteenth-century spare parts and scrap metal, it could still fire the imagination. It even caught the eye of Charles V at the time of his coronation as Holy Roman Emperor in Bologna – one of the greatest ceremonial occasions of the sixteenth century. Both emperor and Pope descended on the city in the Po Valley to perform the solemn ritual bestowing the benediction of God on the political regime of a young man who had inherited the Empire of Spain, along with the fabulous wealth pouring in from its possessions in the New World; the Burgundian lands in what is now Eastern France and the Low Countries; and the central European lands of the

As envisaged by the Enlightenment; the late eighteenth-century statue of Dondi in Padua's vast Prato della Valle swathed in robes recalling a toga, the fingertips of his left hand resting delicately on an armillary sphere, suggests a cloistered intellectual. The reality seems to have been more colourful.

Habsburgs. As a Habsburg, it was of course more than likely that he would be elected Holy Roman Emperor, which he duly was.

Charles V was the universal monarch, master of most of Europe and mysterious lands far beyond the seas, possessed of immense wealth and great personal gifts. Just what could one get the man who had everything the Renaissance world could offer as a present for his coronation? The answer, of course, was the Astrarium components. Rusty and pitted though they had become, they were enough to excite the interest of the most powerful man on Earth.

Charles V summoned craftsmen, astronomers and clockmakers to try to repair the rusted, warped and brittle parts. Only one man said he could repair it, adding, with the confidence of youth, that it would not be worthwhile given its state. Known variously as Gianello Torriano, or Juanelo Turriano, of Cremona, he spent the next twenty years designing a more up-to-date version of the Astrarium and a further three and half years building it, by which time Charles V was no longer the vigorous Renaissance monarch who had been crowned in Bologna by Pope Clement.

Prematurely aged by the state of near constant war with France, rebellious elements of his own dominions, the Ottoman Empire and contumacious Protestant princes, his strength and wealth depleted, he retired to a monastery in Extremadura. Here, with Gianello Torriano entertaining his imperial patron with mechanical automata, including birds that flew and little soldiers who marched, rode, fought and played musical instruments on top of the dining table, the trail of the Astrarium goes cold.

There is a story that the Astrarium was repaired by Torriano and that it accompanied the emperor in retirement, remaining there until it was destroyed, when the monastery was put to the torch by Marshal Soult during the Peninsular War. But when Silvio Bedini, of the Smithsonian, and Francis Maddison, curator of Oxford University's History of Science Museum, published their exhaustive paper on the Astrarium, they were sceptical of this version of events.

Instead, they expressed the hope that it was slumbering in a castle waiting to be awoken from its centuries of sleep. 'It is, perhaps, conceivable that, from the confusion and cluttered storage of some Italian castle there may someday be removed a recognizable fragment of the original Astrarium of Giovanni de'Dondi.'[7]

That was in 1966… and it is still 'perhaps conceivable' that one day history and science will welcome back the rediscovered Astrarium.

1 Quoted in Silvio A. Bedini and Francis R. Maddison, 'Mechanical Universe: The Astrarium of Giovanni de'Dondi', *Transactions of the American Philosophical Society*, Vol. 56, No. 5 (1966), pp. 1–69

2. *Ibid.*

3. De Maisieres [90], Tome XVI, pp. 227–22

4. Silvio A. Bedini and Francis R. Maddison, *op. cit.* pp. 1–69

5. *Ibid.*

6. *Ibid.*

7. *Ibid.*

The Timepiece
Becomes Personal
The Nuremberg 'Pomander' Watch

In 1987, a watchmaker's apprentice was browsing a London flea market, and came across what looked like an interesting box of components, for which he paid £10. Unpacking the box, he found a perforated metal sphere about the size of a billiard ball. Its new owner gingerly unfastened the hooks that latched the hemispheres together and was rewarded with the sight of an assembly of small mechanical components. It was crude by the standards of the late twentieth century, yet half a millennium earlier this metal ball had crystallized the most advanced technology and combined it with the latest fashion. The world's oldest watch had just resurfaced, after an absence of almost 500 years.

The engraving and decoration of the case of Peter Henlein's Pomander Watch was considerably more accurate than its erratic mechanism.

A Nuremberg Egg, one of the earliest personal timepieces, originating from Germany c. 1550, displayed on 17 January 2011 in Geneva, Switzerland at the 21st Salon International de Haute Horlogerie, one of the major watch global exhibitions, where brands premiere their novelties from the 17th until the 21st of January.

Many doubted its authenticity, but after twenty-seven years and some changes in ownership, on 2 December 2014, a panel of experts convening in Nuremberg, where this mechanical object had been made five centuries earlier, decreed that, dating from the early 1500s, it was in fact the world's first watch, and that after forensic examination and laser micro-spectral analysis, microscopic letters had been found engraved on every component.

The longest sequence of letters was 'MDVPHN', which decoded as the Roman numerals for the year 1505, the final 'N' for Nuremberg, while the 'P' and 'H' were taken to be the initials of the maker: Peter Henlein. Indeed, given the number of times that 'PH' had been engraved, the committee believed this to be the watchmaker's own timepiece. It was a historic find.

A statue of Peter Henlein erected in Nuremberg in 1905.

So significant was Henlein to the Third Reich's distorted narrative of German culture that he even warranted a postage stamp.

Watches began to be worn in the late fifteenth century. They developed from spring-driven clocks, which were more compact than weight-driven clocks. However, the personal timepiece was not born out of a need to have the accurate time about one's person: those very early 'watches' were essentially small clocks that could not even be relied upon to deliver accuracy to the nearest hour.

Timekeeping was so appalling that you would have been better off looking at a sundial. These watches did not even have

Long before it became associated with Nazi rallies and war crimes' trials, the city of Nuremberg flowered as a leading creative and intellectual centre of the German Renaissance.

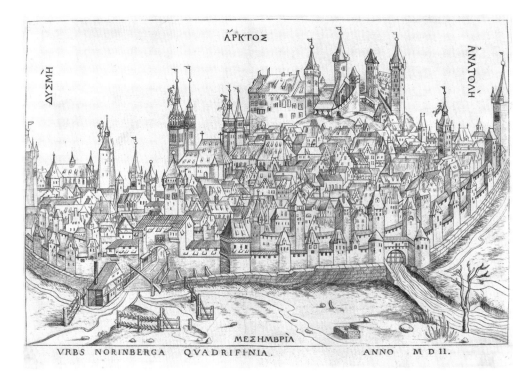

George Heinrich and Kristina Soederbaum in a scene from the film Das unsterbliche Herz, *in which Heinrich played the part of Henlein.*

a hand to record the passing minutes, let alone anything as far-fetched and ambitious as a second hand. Horologically speaking, these early 'drum', 'egg', 'onion' and 'pomander' watches (so named because of their drum-like, spherical or ovoid shapes) were pretty much worthless heaps of scrap metal. However, to their owners, they were a source of immense pride. Slung around the neck on a loop of ribbon or a gold chain, they were worn in a way that aficionados of 1980s hip hop will recall recording artists wearing a logo prised from the grille of a VW or Mercedes.

The primary function of the drum watches that burdened the necks of the fashion leaders of the late Renaissance was to mark their wearers out as different, and to make them feel special. Watches such as that found in a flea market in 1987 would once have been among the most prized of objects 500 years ago; combining beauty, costliness, glamour and the very latest technology.

Before the pomander, Henlein's name had been associated with a watch called the Nuremberg egg, and today Henlein is remembered as the inventor of the watch, in part thanks to the work of Goebbels' favourite filmmaker, Veit Harlan, who showed Henlein inventing the portable personal timepiece in his 1939 film *The Immortal Heart*.

Over four centuries after Nuremberg's watches lent their wearers a sense of that indefinable quality of glamour, they became Nazi propaganda. But the true historical significance of these early watches as the earliest portable mechanical timepieces is as a part of the saga of human development rather than a pawn in a game of nationalist propaganda.

The Howard Hughes *of the* Holy Roman Empire
The Moving Mechanical Musical Boat Clock of Emperor Rudolf II

At an unseen signal, the corps of trumpeters on the galleon's deck raise their instruments to their lips in perfect harmony and an imperial fanfare cleaves the air; drummers, too, begin to play. The sailors and lookouts in the crow's nests stand stiffly to attention, as well they should. They are in the presence of imperial greatness; no less a human being than His Imperial Majesty the Holy Roman Emperor, the most powerful man in Christendom, is on deck.

Seated on his throne under a canopy of cloth of gold, he is serene in contemplation of his great majesty. The fittings and trappings of grandeur surround him, not least the vast double-head golden eagle fixed to the mainmast, the light glinting and darting over its gilded surface.

Standing in front of His Imperial Majesty are seven figures, only slightly less gorgeously dressed. These are the Electors of the Empire, the princes among whom the emperor is *primus inter pares*. Preceded by heralds, one by one they make their obeisance and receive in turn the emperor's blessing.

The ship sets sail, the masts swaying as the vessel rolls and pitches. All the time, the drum beats maintain a solemn dignity broken only by the whistled commands to the sailors. And then the air fills with acrid smoke and the roar of cannon fire.

As the smoke clears, rapturous applause along with cries of delight and appreciation fill the air. This galleon, perfect in every detail, is a ship in miniature. This pageant of imperial power has been played out in scaled-down form for the pre-prandial entertainment of the emperor's dinner guests. The vessel's mainmast 'towers' a metre or so above the dining table, down which it has just trundled under its own power on hidden wheels, music playing on a small internal pipe organ and a tightened drum skin, masts swaying to mimic the movements of a ship under sail.

Everything – the movement, the music, the cannonade – has happened without human intervention, and as the hubbub of appreciation subsides, a crystalline silvery tinkling fills the air. The lookouts atop the masts are striking the sides of the

Sensitive, introspective and fascinated by the occult, Rudolf II may not have been much of an emperor but he was an unparalleled patron of the arts and sciences, he moved the imperial court from its customary seat in Vienna to Prague, which for a few decades became the flame to which the intellectual and artistic moths of Europe flew.

crow's nests with Lilliputian hammers, sounding out the time shown on the clock face positioned just below the imperial eagle.

There is no gong, no bell, no grave 'dinner is served' from a dignified chamberlain. It is instead with this show of magical mechanics from an object that is part clock, part automaton, part Renaissance jukebox, part work of art, that the morose man slumped at the head of the table signifies that it is time to eat.

It is his jaw that one notices first. The beard does little to disguise the expanse of chin, a lower jaw so prominent that it looks like the lip plate of a native Amazonian, distorting the ruff on which the seemingly disembodied head appears to balance. Mandibular prognathism is marked by a distortedly large chin, jutting lower law, swollen lower lip and sometimes unusually large tongue: testifying to a narrow gene pool, it became so identified with one ruling house of Europe that it is still known as the Habsburg jaw.

The man at the head of the table is also at the head of that ruling house. Although he bears but scant resemblance to the little golden figure enthroned on the galleon, he is nevertheless His Imperial Majesty Rudolf II, Holy Roman Emperor. He is a sombre figure; a nimbus of almost ecclesiastical gloom surrounds him, and, chin aside, his most noticeable feature is the pair of dark eyes, deep boreholes of melancholy, looking down a long nose in a gaze that seems unable to decide whether to be disdainful or just weary.

'At the turn of the sixteenth century, the central sensation on the map of Europe came from the sudden rise of the House of Habsburg to a position of immoderate greatness.'[1] By the final quarter of the century, that immoderate greatness was positioned in Prague at the Court of Rudolf II.

Rich in the symbolism of imperial authority, the gilt brass nef prominently displays the double-headed eagle of the Habsburg dynasty over the clock and depicts the Electors paying court to the Holy Roman Emperor.

The Holy Roman Empire over which Rudolf II presided was a vast heterogeneous patchwork quilt of sovereign states that sprawled over modern-day Germany and, at various times, part or all of almost a dozen other modern nations, from the Netherlands to Italy, France to Poland: an imperfectly united European superstate that endured in one form or another for 1000 years. It was said that the imperial crown now sitting uneasily on the highly intellectual brow of Rudolf II had once encircled the head of the great Charlemagne himself.

Rudolf had first worn that crown in 1576, and a leader less like Charlemagne it is difficult to imagine. He showed himself almost incapable of decisive action and was hopeless at uniting the states supposedly under his rule. He did not even fulfil the most basic of dynastic duties, avoiding marriage and with it the production of an heir to give at least the option of future stability (although he did father children by the daughter of the court antiquary). And yet, like Charlemagne, who presided over a flowering of the arts known as the Carolingian Renaissance, Rudolf was the greatest patron of the arts and sciences of the sixteenth century.

The century was defined by its wars: wars of secession, wars of succession, wars of liberation, wars of religion, wars between Catholics and Calvinists, wars between Christians and Muslims, wars that lasted a few weeks, wars that lasted a few decades… In short, Europe in the sixteenth century was boisterous and violent, but Rudolf was not. It was altogether too aggressive for this sensitive, melancholic ruler, so he retreated from it as best he could to pursue his interests in the arts and the occult.

Because he could, he decided to create his own world: a parallel place of beauty and of magic. He moved the imperial court from its customary seat in Vienna to

Opposite: With figures moving, and music playing, a large gilt model of a warship similar to this trundled down the Holy Roman Emperor Rudolf's dining table and let off a Lilliputian salvo of miniature cannon-fire.

Prague, which for a few decades became the flame to which the intellectual and artistic moths of Europe flew: antiquarians and alchemists, poets and painters, musicians and mathematicians all added the lustre of their individual reputations to a city that, by the end of Rudolf's reign, was larger than Paris or London. At the court of Rudolf, one might meet John Dee, advisor to Queen Elizabeth I, for whom magic was as real as mathematics; or Danish astronomer, astrologer, alchemist and aristocrat Tycho Brahe. This was a court so splendid that Rudolf felt it warranted a crown far grander than the one supposedly worn by Charlemagne, and so in 1602 he commissioned another that weighed almost four kilograms.

Rudolfine Prague was a high-minded, big-budget theme park, an imperial Neverland: stuffed with exotic animals – lions, tigers, bears, wolves and flashes of colour from brightly coloured 'Indian crows', as parrots were called at the time. It was peopled by the finest minds of the age, embellished with the most beautiful artworks, and able to demonstrate its sophistication with the most intricately gorgeous and mechanically ingenious objects, among which was the clockwork galleon, or 'nef', with which the emperor amused his guests.

The master of this sort of work was Hans Schlottheim, a citizen of Augsburg, one of the boomtowns of the northern Renaissance. His rise to cultural prominence coincides precisely with the start of the Rudolfine age. In 1576, the year of Rudolf's accession to the imperial throne, Schlottheim, the son of a Saxon clockmaker born in 1547, was named as a master clockmaker within the clockmaker's guild of Augsburg. A decade later, he was appointed a guard or inspector of the guild.

But these waystations on a *curriculum vitae* do not express the sheer ebullient inventiveness of the man: in 1577, he installed a clock on the façade of his house that was the first in the city to strike hours and quarters. City clocks created the societal adhesive of synchronization: places of worship, government institutions, commercial enterprises and private households could regulate their activities by the sound of the bells, and with the increased accuracy of quarterly chiming, the synchronization of those activities became more efficient, as time was subdivided into even smaller useable units.

This was more than a mere civic amenity. In the Europe of the time, major cities were involved in a soft-power game, a sort of cultural and scientific arms race in which clocks were the equivalent of intercontinental ballistic missiles hauled through city centres in the Cold War. While not a clock of the size, splendour and complexity of the fabled Marvel of Strasbourg, Schlottheim's quarter-repeating clock confirmed Augsburg's status as a modern, go-ahead city.

However, this was nothing compared to the work Schlottheim undertook for his royal and imperial patrons: trumpet-playing automata were a speciality, but he could turn

his dexterous hands to, say, a clock embellished with a tableau of automata depicting the Nativity scene, or a pair of clockwork copper crayfish.

Among his later work is a startling four-feet-high clock that depicts the Tower of Babel, albeit one decorated with emperors, gods and personifications of the seven liberal arts: grammar, rhetoric, logic, music, astronomy, geometry and arithmetic. Every minute a rock crystal ball descended outside the tower on a helical track, while a second ball was lifted inside the case. This would trigger automata: Saturn hit a bell with a hammer to accompany the movement of the other gods associated with the various planets. Pipers on a lower level were directly linked to the clock, as was a calendar at the base. Twice a day an organ played a melody to which the musicians raised their instruments. The tower was ornamented with a series of imperial portraits from antiquity to a depiction of Rudolf II, flattering the ruler much in the way that another Renaissance master, Shakespeare, depicted a succession of royalty that linked the Stuart monarchy to the mythical figure of Banquo in the vision vouchsafed Macbeth by the witches.

But Schlottheim's name is most associated with a trio of mechanical galleons, or 'nefs', which have navigated the tempestuous tides of history and survive to this day, one each in the

Jacquemarts in the masts sounded out the hours, striking the sides of their metal 'crow's nests'.

The mechanical engine that powered an object that is part clock, part automaton, part Renaissance jukebox, part work of art.

British Museum, Vienna's Kunsthistorisches Museum and France's National Museum of the Renaissance.

They are spectacular works of art, drawing on the skills of enamellers, clockmakers, jewellers, painters, sculptors, naval architects, gunmakers and musicians. Schlottheim's skill was not just that of a gifted clockmaker, capable of devising a system of wheels, springs, pinions and arbors for the mechanical replication of human movement, the motion of a vessel at sea, the activation of a firearm and of course the telling of time, but also that of a studio master. Today such work would be called project management: coordinating the skilled artisans expert in various disparate fields, whether gilders or musical-instrument makers.

Schlottheim was the conductor of an orchestra of talent before knowledge and skill became diversified and the era of the specialist dawned. This was still the age of the Renaissance man, an ideal that did not merely exist among courtiers and rulers, but also found its expression in men such as Schlottheim, who were skilled technicians, artists, mechanics and musicians capable of combining their skills and the abilities of others to create objects that brought the sum of human knowledge to the boundary of magic.

Clocks were just such objects. Sixteenth-century status symbols par excellence, they bolstered the illusion that man had domesticated the terrible predator time, and, as time was precious, clocks became a medium through which the practitioners of the applied arts could express the importance of the object's owners. The 'nef' represented the apogee of this type of sophisticated display.

But this was more than just a fashionable gadget. As one of his biographers explains, Rudolf was obsessed 'with the belief that man-made instruments could extend the limits of the human senses'.[2] And among man-made instruments, clocks were especially dear to Rudolf: they were powerful allegorical objects in which science and magic met.

The galleon clock was stuffed with metaphor and meaning: as a 'ship of state', it functioned perfectly, running like the clockwork it was. The cannons were a warning to those who might dare question His Imperial Majesty, and obviously the pantomime of submission by the electors stressed the suzerainty of the emperor.

Julia Fritsch, the curator of a 1999 exhibition that united the three nefs, suggested that Rudolf commissioned the galleons as items of tribute rendered to the Sultan in the fifty years of peace between the Holy Roman and Ottoman Empires.[3] However, a mention of one 'nef' in an inventory of the collection of the Elector of Saxony (a document located in 2004) suggests that it might have been a diplomatic gift *within* the empire to gain the favour of an influential elector (while simultaneously reminding him of his place).

If indeed these were the intended purposes, then they are poignantly ironic, as they guaranteed neither harmony with the Ottomans nor the authority of the emperor within his realm.

Rudolf was a fascinating, thoughtful character and one of history's great eccentrics, but he lacked the qualities of a Renaissance leader, and when he finally did go to war against the Ottomans, it was such a disaster that his brother carried out a coup and placed him under house arrest. Rudolf is said to have tried to invoke magical forces to punish his brother, but, alas, they did not appear at his bidding and instead it was he who died in Prague.

---

1. Norman Davies, *Europe: A History*, p. 524
2. Peter Marshall, *The Mercurial Emperor* (London: Pimlico, 2007), p. 99
3. Julia Fritsch, *Ces curieux navires: Trois automates de la Renaissance* (Paris: Réunion des Musées Nationaux, 1999), p. 19

Buried Treasure
The Emerald Watch of the Cheapside Hoard

By June 1912, all that was left of the buildings that had stood on the corner of Friday Street and Cheapside for 250 years was a pile of rubble. The new buildings that would rise in their place required deeper foundations than before, so labourers were now at work on the cellars, hewing at the floor with pickaxes, removing the last vestiges of the seventeenth century from the site. Then one of them spotted something glinting in the earth. His pick had splintered the top of a rotten wooden casket and opened a window to the past, to Jacobean London.

Hurriedly the flooring and remains of the casket were cleared away. The labourers could not believe their luck; life as a navvy in the early twentieth century was not an enviable one, but these men had inadvertently uncovered the

The star of the Cheapside Hoard – a pocket watch installed within a large Colombian emerald.

most significant collection of seventeenth-century jewels ever discovered. In all there were hundreds of pieces: tiny hardstone intaglios the size of a farthing; long looping sautoirs; and brooches and pendants all tangled together and embedded in great lumps of London clay.

The buildings were the property of the Worshipful Company of Goldsmiths, one of the most prestigious of London livery companies, and the contract for their demolition had included a clause that stated the lessees could take what they liked in terms of architectural salvage – 'save and except any antiquities and articles or objects of interest or value which shall be preserved by the lessees and handed over to the lessors'.[1]

However, what the lessors and the lessees might have agreed between themselves was of little interest to men whose hands were callused from a lifetime's use of picks and shovels, rather than stained with the ink of legal documents. It could be that these navvies did not even know that they had discovered buried treasure; one of them said he thought they had dug up an old toy shop. They knew they had found something, though, and they also knew exactly where to take it.

'It is perhaps the strangest shop in London. The shop sign over the door is a weather-worn Ka-figure from an Egyptian tomb, now split and worn by the winds of nearly forty winters. The windows are full of an astonishing jumble of objects. Every historic period rubs shoulders in them. Ancient Egyptian bowls lie next to Japanese sword guards and Elizabethan pots contain Saxon brooches, flint arrowheads or Roman coins.'[2]

So wrote H. V. Morton of a real curiosity shop at 7 West Hill in Wandsworth, owned by George Fabian Lawrence. Lawrence, or Stoney Jack as he was known to many of the thousands of labourers who toiled on Edwardian London's numerous building sites, was short, wheezy, moustachioed and bespectacled, with a weakness for cheap cigars. In sum, he looked like a genial bank clerk, but although he did not enjoy the celebrity of Howard Carter, he was nevertheless one of the great archaeologists of the age. For almost fifty years, from the 1890s until his death in 1939, he pursued a one-man archaeological investigation of London.

At the time, the city was undergoing significant redevelopment and expansion. As most of the heavy work was still being done with picks and shovels, items from London's past were constantly being uncovered by navvies. Lawrence toured the capital's building sites striking up conversations with labourers, making it known that he would buy anything they turned up, whether it was a single arrowhead or, as in this case, a lump of metal-studded clay.

Once Stoney Jack had washed away this clay, he saw precious stones that glittered once again in light they had not seen for over a quarter of a millennium. He was looking at the biggest find of his career.

View of Cheapside showing the entry of Queen Marie de Medici, mother of Louis XII and wife of Henry IV of France, into London on 29 October 1637.

Part 'fence', part 'antiquary', and later member of the staff of the Guildhall and London Museums, Stoney Jack was the unofficial conduit for objects between the building sites of London and the city's major museums. He operated in a crepuscular area that was neither entirely legitimate nor overtly criminal. Archaeology was often the preserve of amateurs, and museum curators were less fettered than today by compliance and regulation when it came to acquiring items. Sometimes he would buy smaller items on the spot, but larger hauls and more significant finds that might otherwise attract the attention of building site owners and foremen would be smuggled out and brought to West Hill on Saturdays. Lawrence's motives were not, overly, mercenary. He was known to be fair, both by the navvies who sold to him and the museums who willingly bought from him: towards the end of his life he claimed to have 'got 15,000 objects out of the soil of London in 15 years for the London

CHRESTIEN DANS LA VILLE DE LONDRES.

Museum alone'.[3] His chief motivation was a genuine love of the past, captured in the objects preserved beneath London's buildings and pavements.

He believed that the as-yet-unopened London Museum should have the hoard, and on the same day the trustees met to discuss its acquisition, Lawrence was appointed inspector of excavations. The existence of the hoard was kept secret for two years until the opening of the new London Museum at Stafford House, by King George and Queen Mary in March 1914.

'One of the principal attractions', reported *The Times*, 'is the Gold and Silver Room.'

> *It contains a unique survival of the early part of the 17th century in the shape of a collection of jewelry discovered in the City. This treasure was found buried in a box, and was part of a jeweller's stock. There are many duplicates, and some of the articles are in an unfinished condition. Altogether 340 pieces were discovered, including rings, pendants, chains, scent-bottles, pomanders and watches, and part of a Communion set in crystal and gold. The delicacy and*

elegance of the designs and workmanship are remarkable, and in one or two instances it is curious to recognize the resemblance of the ornaments fashioned in Jacobean times to the art nouveau of the present day.[4]

Among them all, one piece in particular stood out: an emerald the size of an egg that had been cut and hinged to enable its wearer to wind the watch set inside.

Dating from around 1600, 'The emerald-cased watch is not just the most spectacular item in the Cheapside Hoard'. According to Museum of London curator Hazel Forsyth, it is 'one of the most remarkable jewels in the world'.[5] The watch itself is a miniature masterpiece with its engraved dial, fusee and three-wheel gear train – not, of course, that it would have been particularly accurate. But that was not the point.

King James I, painted here by John de Critz the Elder (c. 1552–1642), presided over a court in which men and women alike were bedizened with jewels.

This was a rich man's toy, a status-conferring object that was remarkable even by the standards of personal decoration that prevailed at the court of King James I, where the acknowledged style leader George Villiers, Duke of Buckingham, would appear at 'any ordinary dancing' in 'clothes trimmed with great diamond buttons and to have diamond hatbands, cockades and earrings, to be yoked with great manifold knots of pearl, in short to be manacled, fettered and imprisoned in jewels'.[6]

As well as speaking eloquently of the decadent splendour of the Jacobean court, however, this sparkling survivor from the seventeenth century tells of a time when the world was expanding, and London was emerging as one of the great mercantile cities of the world. 'The emeralds used to form the case and lid come from Muzo and were almost certainly designated "stones of account"', but while the famous Muzo mine was in South America, this stone may have arrived in Britain from the East, because thousands of Colombian emeralds were supplied to India and Burma to satisfy local demand. The Burmese were so keen to obtain emeralds that they bartered rubies for South American stones, and in Goa in the 1580s emeralds were valued 'as muche as a Dyamond and somewhat more'. European dealers were also keen to buy emeralds in Eastern markets so that they could be labelled 'oriental' stones, because oriental stones fetched a premium price in the Low Countries of France, Germany and England.[7]

This remarkable time-telling jewel dates from a pivotal point in the growing commercial importance and global power of England. Having vanquished the Armada and established colonies in northern America, England looked hungrily to other corners of the world, and in 1600 an ageing Elizabeth I granted a Royal Charter to the East India Company giving the green light to almost 300 years of state-sanctioned commercial opportunism, including slave trading and opium smuggling, that resulted in national and individual enrichment… all the while laying the foundations of the future empire.

Beyond the obvious purpose of keeping it secure, no one knows why 'the Cheapside Hoard' was buried. Was its owner going overseas? And if so, why did he not return? Was there some existential hazard such as war, plague or fire that caused the jeweller to bury his stock, only to perish himself in said disaster? Similarly, one can only speculate as to how it arrived in seventeenth-century London, and as to its manufacture, even Hazel Forsyth is baffled.

'It is possible', she says, 'that the watch emeralds were cut in a lapidary workshop in Seville or Lisbon or conveyed by some other route to Geneva, a city renowned for the skill of its watch and hardstone case makers.'

During the Reformation of the mid-sixteenth century, decrees from the Protestant leader Jean Calvin in Geneva restricted the level of permissible personal ornament,

A street of wooden miners' cottages in Muzo, Colombia, the world's emerald capital.

anti-luxury sumptuary laws were enacted, and goldsmiths were forced to turn to making cases for watches, beginning the city's reputation as a centre of horology.

It is also possible that the cutting of the stones took place in London. All we are sure of is that it was one of the most extravagant and costly items of the age in which it was made. As Hazel Forsyth says: 'What is not in doubt is that the lapidary skills of the case maker are of the highest order, demonstrating a technical mastery of the material.'[8]

Its mystery makes it all the more compelling. Its haunting beauty and the absence of any hard facts beyond the colourful nature of its discovery encourage the mind to wander across the world as it was in the early 1600s: from the newly discovered Muzo mine of Colombia, across the pirate-infested high seas to the gem markets of India and Burma, then on to Geneva, the Low Countries or one of the great cities of imperial Spain, before finding itself on Cheapside – the bustling thoroughfare that lay at the heart of Tudor and Stuart England with its merchants and whores, goldsmiths and gaudy aristocrats. It was this lively cross-section of life that inspired Thomas Middleton to write his play *A Chaste Maid in Cheapside* – a bawdy comedy about a rich Cheapside jeweller trying to marry his daughter into the lower echelons of the nobility… just such a man as might have buried these jewels in his cellar for safekeeping.

1. Quoted in Hazel Forsyth, *London's Lost Jewels* (London: Philip Wilson Publishers, 2013), published on occasion of the exhibition *The Cheapside Hoard: London's Lost Jewels*, Museum of London (11 October 2013–27 April 2014)
2. H. V. Morton quoted on Smithsonian.com
3. Quoted in Hazel Forsyth, *London's Lost Jewels, op. cit.*
4. 'Romance and Colour of London', *The Times*, 19 March 1914, p. 6
5. Hazel Forsyth, *London's Lost Jewels, op. cit.*
6. MS in the Harleian Library, quoted in James Robinson Planche, *History of British Costume* (London: Charles Knight, 1834), p. 275
7. Hazel Forsyth, *London's Lost Jewels, op. cit.*
8. *Ibid.*

A
CHAST MAYD
IN
CHEAPE-SIDE.

A
Pleasant conceited Comedy
neuer before printed.

As it hath beene often acted at the
Swan on the Banke-side, by the
Lady ELIZABETH her
Seruants.

By THOMAS MIDELTON Gent.

LONDON,
Printed for *Francis Constable* dwelling at the
signe of the *Crane* in *Pauls*
Church-yard.
1630.

Thomas Middleton's play A Chaste Maid in Cheapside *is a bawdy comedy about a rich Cheapside jeweller trying to marry his daughter into the lower echelons of the nobility… just such a man as might have buried the jewels that would become the 'Cheapside Hoard' in his cellar.*

Japan's Moveable Hours
Wadokei

As a harvester of souls, St. Francis Xavier travelled to the ends of the Earth. He had brought the word of God to India, Indonesia, Malaysia, Timor and now, on 15 August 1549, the Feast of the Assumption, he arrived in the Japanese port of Kagoshima. The 43-year-old priest was the greatest missionary the Catholic Church would ever know. A decade and a half earlier, he and six others had taken vows of chastity and poverty in a chapel in Montmartre. These seven called themselves the Society of the Friends of Jesus, but would become known by a less cumbersome name: Jesuits.

Francis Xavier was a founder of one of the most influential religious orders, and its early success owes much to his zeal and energy.

Harvester of souls extraordinaire, Saint Francis Xavier was one of the seven founding Jesuits and carried the order's teachings across the world's oceans to India, Borneo, and many other far-flung lands including Japan, where he is also responsible for the introduction of the mechanical clock. (Painting attributed to seventeenth century Spanish School.)

Intellectual and spiritual exchange: Francis Xavier the Jesuit missionary, talking to Buddhist monks.

Europeans had only recently arrived in Japan – and by accident at that. Six years earlier two Portuguese merchants blown off course during a storm had taken shelter in a cove on Tanegashima island. But even though it was a land distant from Rome, veiled by mystery, this well-travelled Jesuit sensed a rich crop of converts, and with his arrival began the period known to historians of Japan as the Christian Century.

The spirit of Christian brotherhood, peace and goodwill was conspicuous by its absence in mid-sixteenth-century Japan. Francis Xavier had made landfall in a country in a state of near constant conflict between *daimyos* (regional warlords). There may have been an emperor in Kyoto, but effective power lay with men such as Ōuchi Yoshitaka, who ruled his fiefdom from the city of Yamaguchi on the southern tip of Japan's main island.

Frustrated that Christianity was not getting what today would be called traction, and having been denied an audience with the emperor, Xavier decided to try his local warlord. His initial visit to Yamaguchi had been made in a spirit of Christian meekness. It had not been a success. The next time he determined to do things differently, and in 1551 a much-changed Francis Xavier presented himself before Ouchi Yoshitaka dressed like an ambassador, and bringing ambassadorial gifts. He was rewarded with permission to practise his religion and make converts, even given an empty Buddhist monastery to use.

While Xavier preached and converted, Yoshitaka enjoyed the presents given to him by his new friend, which just so happened to include one of the great technological marvels of Renaissance Europe: a clock.

There was not much time for Yoshitaka to enjoy his clock and Xavier to convert his infidels: the former was overthrown and killed later that year, and the latter died in December 1552. However, the teachings of Christ and the mechanical timepiece had taken root in Japan, more Christian priests followed, and in 1600 a school was established by missionaries in Nagasaki that taught, among other things, clockmaking.

Then, in 1603, a shōgun came to power in Edo, modern Tokyo, initiating an eponymous period of feudalism, also called the Tokugawa shogunate, that amounted to a quarter of a millennium of national isolation. During the second decade of the seventeenth century, as warring states made peace and submitted to Tokugawa, Christian missions were closed, and during the 1630s Christianity was banned altogether. By 1639, the policy known as *Sakoku* (closed country) had taken effect: foreigners were expelled, the Japanese were prevented from travelling overseas, and almost all foreign trade was brought to an end.

One of the few remaining traces of European culture to remain, and indeed flourish, in Edo-period Japan was clock-making. Edo, the city from which Japan was ruled,

A clock-maker of Old Japan.

also became the country's horological capital, and for the next 250 years Japanese clockmaking would take an evolutionary path that diverged from the way it advanced in Europe, creating a fascinating alternative horological culture.

The clocks introduced to Japan by Francisco and other Europeans adhered to the notion of time as fixed – a conceptual grid to give life order. However, according to social anthropologist and Professor Joy Hendry, who has made Japanese life and customs the subject of decades of study, the Japanese concept is somewhat less rigid: 'The indigenous Japanese word for time stands very much for a point in time, a particular moment or occasion, rather than an abstract continuing entity. There is even a sense that time can be "folded" or "manipulated" according to ecological or social needs.'[1]

And once the Catholics were kicked out, taking the inelastic notion of time with them, master clockmakers such as Sukezaemon Tsuda were free to develop a timepiece that met the complex and fluctuating nature of time as understood by Japan during that period. It was called the Wadokei, and was wholly peculiar to the nation that invented it.

'Wadokei were extremely complex because of the nature of the Japanese system of time, which used the lunar calendar,'[2] writes John Goodall in his history of Seiko watches. Clocks were wound and set at the beginning of that Japanese day, which commenced at dusk. Significantly, the temporal day did not begin at a set time, such as midnight, but when daylight ended. Dictated by sunrise and sunset, each day was divided into day and night, each comprising six periods of time known as *koku*. These *koku* varied in duration, with seasonal changes in the length of day and night.

Japanese clockmakers had devised a system whereby the regularity of the Western clockwork could be adapted to the specific need of dividing each two periods of irregular times into six periods of constantly varying but identical duration. Thus, at the summer solstice, each of the six nocturnal *koku* would be at their shortest, the diurnal at their longest; and vice versa at the winter solstice. Using the foliot balance, time either was slowed down or sped up by moving the regulating weights at the end of the foliot arm further apart or closer together. This altered the speed at which the clock ran, to accommodate the different lengths of night. However, as the length of nocturnal and diurnal *koku* varied slightly every day, the weights had to be moved twice daily.

As the Edo period progressed, more sophisticated timepieces appeared with alarms and carillons, and then came a development unseen anywhere else in the world: at the end of the seventeenth century, a 'double foliot balance' clock was developed. [3]

There is a lantern clock in the permanent collection of the Seiko Museum which dates from 1688 and is attributed to Skezaemon Tsuda. At first glance, it is not unlike European lantern clocks of the period, but it is of immense historic significance, as it is 'the oldest available clock with a double foliot balance' [4] capable of switching automatically between daytime and nighttime. This cut the number of human interventions needed to address the constantly shifting lengths of the *koku*.

Another system of coping with this shifting pattern of temporal units was devised later in the Edo period, when the foliot balance was superseded by pendulum or spring regulators, which were more reliable but more difficult to speed up or slow down as the seasons demanded. The simple and elegant solution of a rail around the dial was introduced, along which the hour markers could be moved further apart or closer together.

To add a further Japanese twist, the twelve *koku* were not numbered, but named according to the Japanese zodiac. There were just two fixed times: horse o'clock (midday) and rat o'clock (midnight); the remainder of the astrological menagerie – dragon, ox and monkey, *inter*

Dating from the early Edo period and ascribed to Sukezaemon Tsuda (the third). This is the oldest extant clock with a double foliot balance. It is designed to automatically switch between daytime and night time at dawn and dusk using the two foliot balances.

A Western clock face on a wooden box in a Japanese setting. By Ryuryukyo Shinsai (c. 1764–1820).

alia – moved around the dial, bunching up or spreading out as the length of daylight dictated. So, sunrise took place at 'rabbit o'clock' and, perhaps just to underline the divergence from Western timekeeping, 'rooster o'clock' was the hour of sunset.

By the middle of the nineteenth century, the Japanese timepiece had evolved into a device that looked nothing like its counterparts in the Western world. The pillar clock showed time by means of an indicator that moved between the top and bottom of a long dial occupying the majority of a timepiece that looked like a wall-mounted barometer.

And then, on 1 January 1873, Wadokei clocks suffered a mass extinction event. At the end of the 1860s, after a quarter of a millennium sequestered from the rest of the world, the shogunate was overthrown and the Meiji emperor restored.

Although it spanned just a few decades (from 1868 until 1912), the Meiji period reversed Edo's isolationist policies and saw the country embark on a course of crash modernization. The world had changed in the three centuries since Francisco's arrival in Japan; the age of exploration had rapidly become the age of colonisation, and emerging from the Edo period the country faced the shameful prospect of becoming a colony of one or other of the modern industrially developed military powers – a prospect that could be avoided if Japan itself became a modern, industrially developed military power. This was a process, however, that would require joining the time-reckoning system used by the rest of the world.

In 1872, the fifth year of the Meiji period, an imperial edict was issued, replacing the traditional Japanese calendar with the Western, solar calendar. It ordered that, along with the calendar, the method of telling the time also needed to be aligned with the international standard.

This [new] system, which has helped our country's government in all spheres since the establishment of international ties with other lands, must be adopted… We hereby order that the solar calendar be published throughout the land, using an equal number of hours to divide day and night, thereby reforming the time system, not only in order thereby to put right the calendar law, but to promote the enlightenment of the people, as well…

Whereas, until now the system of hours has been comprised of a rendering into twelve hours varying in duration in accordance with the length of day and night, we hereby newly establish a system of time comprised of twenty-four hours equal in duration for both night and day, dividing the period of time lasting from the hour of the rat until the hour of the horse into twelve hours, and naming this period 'forenoon,' and dividing the period of time lasting from the hour of the horse until the hour of the rat into twelve hours, and naming this period 'afternoon.' [5]

The regime was in no doubt that changing the way the nation calculated the time and date was crucial: 'in order to modernize our country and reform the ancient customs with the goal of moving forward into the realm of civilization as a people, it is of utmost urgency that we hereby rectify the calendar law.'[6]

Certainly, by 1905, with the defeat of Russia by a modern Japanese navy, it was impossible to deny that Japan had emerged as a twentieth-century world power. But this remarkable achievement had not been without cost: many traditional aspects of Japanese life had needed to be sacrificed, among them the Wadokei.

Opposite: Dating from the late Edo period, this wall clock with 'Circle Graph Dial' has a timescale like a pie chart, calibrated for the seasonal time system. Representing the ultimate technological evolution of Japanese watchmaking over many generations of isolation from the rest of the world; the hand automatically extends and retracts from season to season to indicate the seasonal times (the hands become longest on summer solstice and shortest on winter solstice).

Below: 'Illustration of the Imperial Diet of Japan' by Gotÿ Yoshikage, 1890. Although it spanned just a few decades (from 1868 until 1912), the Meiji period reversed Edo's isolationist policies and saw the country embark on a course of crash modernization: this included adopting the Western notion of time.

1. Joy Hendry, 'Time in a Japanese Context', essay in exhibition catalogue 'The Story of Time' (London: Merrell Hoberton, in association with National Maritime Museum, 1999)
2. John Goodall, *A Journey in Time: The Remarkable Story of Seiko* (Herts, UK: Good Impressions, 2003), p. 8
3. https://museum.seiko.co.jp/en/knowledge/wadokei/variety/
4. *Ibid.*
5. Hoshimi Uchida, 'The Spread of Timepieces in the Meiji Period', *Japan Review*, No. 14 (2002), pp. 173–192
6. *Ibid.*

The Search *for* Longitude
Harrison's Marine Chronometer

It was October 1707, and a flotilla of victorious warships was returning to Britain after a successful summer spent harrying French shipping in the Mediterranean. The British ships were under the command of the celebrated Admiral Sir Cloudesley Shovell, who had begun his naval career as a 13-year-old cabin boy and who was now, forty-four years later, a portly man with small, hard eyes, multiple chins that could not be hidden by his neckcloth and, most importantly, a formidable reputation. After almost half a century at sea, he knew the perils of the fog and gales that had hampered their journey for almost a fortnight, but his navigators' calculations told him that he was safely in the mouth of the English Channel.

Now a prized part of the National Maritime Museum, and regarded as one of horological history's great milestones, H1, John Harrison's first attempt to the solve the longitude problem, extended the limits of the technology of its day and dazzled scientists and horologists alike.

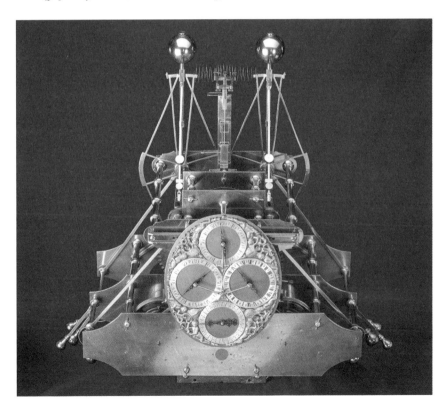

As well as a valorous naval officer and an experienced sailor, according to one account he would appear to have been a strict disciplinarian who dealt fiercely with insubordination. It is said that a crew member who had suggested the Admiral's calculations were incorrect had been hanged on the spot for mutiny. It was a harsh penalty; harsh and, ultimately, unjust, as suddenly rocks loomed out of the fog where there should have been none.

The Scilly Isles had appeared right in the path of the Admiral's ships.

Four vessels sank, almost 2000 lives were lost (more than the entire British dead and wounded at Trafalgar a century later), including the Admiral (although one version of events tells that he made it ashore alive, where he was promptly murdered for an emerald ring he wore).[1]

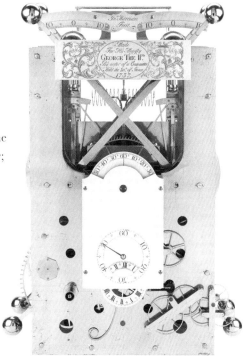

John Harrison's second marine timekeeper, H2, built on the successes of its predecessor but lacked its flamboyant appearance.

The disaster and the huge loss of life, not through battle but as a consequence of inadequate navigation, was a profound shock to the nation. All the more profound because it was far from isolated: every few years around the turn of the seventeenth century there were major maritime disasters: in 1691, a number of battleships had been lost off Plymouth; in 1694, a squadron led by Admiral Wheeler crashed into Gibraltar believing they had already passed it; and in 1711, more boats were lost near the St Lawrence River, having veered fifteen leagues off course in just one day.[2] All were caused by navigational errors.

By the second decade of the 1700s, there was increasing public pressure for the government to do something. But what?

'Several Captains of Her Majesty's Ships, Merchants of London, and Commanders of Merchantmen, in behalf of themselves, and all others concerned in the Navigation of Great Britain' were in no doubt as to what was needed. On 23 May 1714, they lodged a petition at the House of Commons stating that 'discovery of the longitude' was a matter of immense 'consequence to Great Britain', and that:

for want thereof, many Ships have been retarded in their Voyages, and many lost; but if due Encouragement were proposed by the Publick, for such as shall discover the Same, some Persons would offer themselves to prove the same, before the most proper Judges, in order to their entire Satisfaction, for the Safety of Men's lives, Her Majesty's Navy, the Increase of Trade, and the Shipping of these Islands, and the lasting honour of the British Nation.[3]

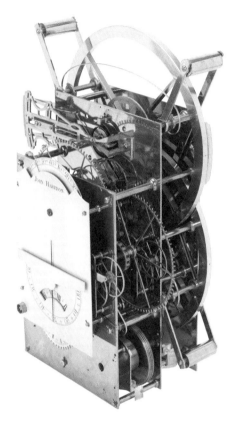

The age of exploration was at its height: a centuries-long game of supermarket sweep with the nations of Europe racing around the world to accumulate territories overseas. The nation that commanded the seas, commanded the world. Britain's very future was at stake.

Most, if not all, of the major losses of life in shipping could have been avoided had the ships been able to determine their longitude. Latitude could be determined by taking astronomical

Harrison's third attempt to solve the longitude problem.

readings, but this was only half the information needed. Indeed, some of Sir Cloudesley's ships had taken latitude readings and disaster had still struck. To accurately determine the location, the latitude needed to be cross-referenced with longitude, and at the time Sir Cloudesley met his end, mariners used the notoriously unreliable system of dead reckoning, the key element of which was dropping a piece of wood over the side and seeing how long it took the ship to pass it, to give an approximate longitude.

Discovering the longitude was the compelling scientific issue of the day – much as, say, nuclear weapons and spacecraft would obsess the scientists of the Cold War. But, unlike nuclear weapons and the space race, there were direct and readily comprehensible benefits to being able to determine longitude: lives would be saved, trade would be increased and Britain would become great. It was a journey into the scientific unknown, and some truly bizarre solutions were considered, including a scheme for mooring vessels equipped with flares to be fired every night at midnight along the major trade roots – a scheme so truly bonkers and splendidly impracticable that its serious consideration indicates the level of desperation.

John Harrison, English inventor and horologist, is shown seated beside a table, holding the watch made to his design in 1767. (Oil painting by Thomas King.)

It became a government priority. Those with long memories could be forgiven for experiencing a feeling of déjà vu. Forty years earlier, on 15 December 1674, back in the days of King Charles II, whose merchant fleet was the largest in the world, a royal commission had been appointed to solve 'the longitude problem'. A year later, the monarch's continuing close interest in matters of navigation led to the foundation of the Royal Observatory at Greenwich, 'so as to find the so-much-desired longitude of places for perfecting the art of navigation'.[4]

Nevertheless, in 1714, a parliamentary committee was formed in front of which expert witnesses appeared, including, on 11 June, the president of the Royal Society, the defining genius of the age, Sir Isaac Newton. Sir Isaac informed committee members of the difficulties they faced. Disposing of the mid-ocean flare system, he told them there remained three other alternatives:

> *One is, by a Watch to keep Time exactly: But, by reason of the Motion of a Ship, the Variation of Heat and Cold, Wet and Dry, and the difference of Gravity in different Latitudes, such a Watch hath not yet been made.*
>
> *Another is, by the Eclipses of Jupiter's Satellites: But, by reason of the Length of Telescopes requisite to observe them, and the Motion of a Ship at Sea, those Eclipses cannot yet be there observed.*
>
> *A Third is, by the Place of the Moon: But her Theory is not yet exact enough for this Purpose: It is exact enough to determine her Longitude within Two or Three Degrees, but not within a Degree.*[5]

In short, Newton's testimony before the committee in search of longitude can be summed up in two words: 'Good luck'.

Just in case they were minded to pursue this fool's errand, they would need watches, Newton told them: 'In the three first Ways there must be a Watch regulated by a

Spring, and rectified every visible Sun-rise and Sun-set, to tell the Hour of the Day, or Night'. Additionally, 'In the first Way, there must be Two Watches; this, and the other mentioned above',[6] i.e. the watch that 'hath not yet been made'.

The committee was left in no doubt that, in order to find longitude, they would first need to find a way of keeping time at sea.

Establishing longitude requires knowledge of two times: that aboard ship, usually set at the local solar noon, and that at a known longitude, for example the port of departure. With the knowledge of the time simultaneously in two places, the difference between the two can be converted into a distance from the reference point. The mathematics is hardly Newtonian in complexity: one full 360-degree rotation of the Earth takes twenty-four hours; thus, one hour's difference converts to 15 degrees of longitude.

However, if the maths were simple, the clockwork was complex. As Newton had helpfully pointed out, no such clock existed that was capable of remaining impervious to accuracy-altering assaults made on it by violent motion, not to mention the effect of heat and cold on the expansion and contraction of metal and the viscosity of lubricating oil.

Their avenues exhausted, the committee resorted to one of the great stimulants of ingenuity: greed. On 20 July 1714, a dying Queen Anne gave her assent to the Longitude Act, which held out the promise of extraordinary riches to the man who solved the scientific problem that had perplexed and confounded generations of mariners and natural philosophers, as scientists were known at the time.

A prize of £10,000 was offered to the inventor of a system accurate to within one degree; £15,000 would be awarded if accurate to within two-thirds of a degree; and the grand prize of £20,000 would be released if longitude could be determined to within half a degree. A degree equates to approximately sixty-nine miles at the equator, tapering to nothing at the poles. By twenty-first-century standards, accustomed to location accuracy to within a metre or two available on mobile telephones and satellite navigation systems, these tolerances may seem enormous, but in the seventeenth century they were believed to be impossible to achieve.

The terms of the award were careful to stress that, in order to claim the prize, the method needed to be tested at sea and deemed 'practicable and useful'. Nevertheless, this vague caveat in the 'small print' did little to discourage submissions. The lure of £20,000 (approximately £3,750,000 in 2018[7]) attracted cranks and mountebanks, as well as clockmakers and inventors who deluged the Board of Longitude – a jury comprising prominent mathematicians, astronomers and expert navigators – with proposals. Even the octogenarian Sir Christopher Wren was tempted to try his

hand and suggested a trio of devices, encoding his suggestions for security.

Soon there was not a town in the kingdom that was unaware of the race to discover longitude and find the glittering crock of £20,000. It was a sum unimaginable to most men; an artisan, a skilled carpenter, could expect to earn £40 per annum. But John Harrison of Lincolnshire, who celebrated his twenty-first birthday in the same year as the Longitude Act was passed, was more than just a skilled carpenter. A ferocious autodidact, he had devoured scientific texts lent to him by a clergyman, and before he was twenty, he had built his first clock… making it entirely out of wood.

His expertise grew, and with it his reputation. By the time he was in the his late thirties, he was ready. In 1730, he went to London, as many have before and since, to seek his fortune. For the remainder of his long life – and he would live for almost another fifty years – his fortunes remained inextricably tied to the search for longitude.

By then longitude had become a popular euphemism for an impossible task. In his 1726 novel *Gulliver's Travels*, Jonathan Swift portrayed its discovery as a utopian chimera – as fanciful as perpetual motion and universal medicine. Indeed, by 1730 many believed that the prize would remain unclaimed because longitude would never be accurately determined. It would appear that the Board of Longitude was among their number: 'Although that august body had been in existence for more than fifteen years, it occupied no official headquarters. In fact, it had never met.'[8] However, Harrison knew he would find one of its most prominent members in Greenwich.

Edmund Halley had succeeded John Flamsteed as Astronomer Royal, and he received Harrison warmly. He heard the country carpenter's ideas and found them not without interest. However, Halley was a stargazer and not a clockmaker, so he suggested that Harrison see George Graham, the age's most celebrated

Shown here with his son John Roger Arnold and his wife, examining a chronometer, John Arnold was one of the generation of British watchmakers who adopted Harrison's principles and put marine chronometers into series production.

horologist. A Fellow of the Royal Society, Graham was far more than a maker of fashionable timekeepers; he made hi-tech objects that were pushing the boundaries of knowledge and illuminating the darkest recesses of ignorance. Harrison was in the presence of greatness, but there must have been something intriguing about the carpenter-turned-clockmaker as the two men met in the morning and talked until dusk. Impressed, Graham loaned him some money and raised further development funding from the East India Company.

Harrison returned home to work on the clock that is today known as H1. It took five years to complete, and to the twenty-first-century eye looks less like a clock and closer to some sort of steampunk lunar landing module assembled according to plans drawn up by Heath Robinson. Three feet high, made of shining brass with rods, wires, buttons and spheres protruding at strange angles, it weighed a considerable 75 lbs. It is the sort of object that would look right at home in the laboratory of a black-and-white-filmed adaptation of *Frankenstein*. It was like nothing anyone had seen, but Halley and Graham were commendably open-minded, and a sea trial was arranged. It was not the Caribbean voyage that the terms of the competition specified, but a round trip to Lisbon, during which H1 showed far greater indifference to the stormy conditions than the seasick Harrison.

So successful was H1 that the Board of Longitude staged that rarity of rarities: a meeting – some say the first since its founding. It was a convocation of the finest minds in the country: professors from Cambridge and Oxford, top naval officers, Graham, Halley and that great intellectual magpie Sir Hans Sloane, president of the Royal Society, among them. They were astonished at H1's accuracy, and a more opportunistic man would have pushed for a definitive sea trial to the West Indies, but Harrison felt that he could make further improvements and instead sought a grant for the 86 lb H2. Slightly more conventional-looking and taller than H1 (which subsequently went on display in Graham's shop and drew visitors from all over Europe), H2 was still very much a child of its idiosyncratic parent. Its rectangular brass structure looks curiously timeless, and, but for the elaborate scroll engraving on a name plate dedicating the machine to George II, it would be hard, at a glance, to date.

H2 took two years to build, but by the time it was complete its perfectionist maker believed it to be out of date, and thereafter, barring occasional requests for further funding, Harrison disappeared into his workshop and embarked on a near two-decade mechanical odyssey to create H3, which, when complete, was immediately superseded by a saucer-sized 3lb watch that Harrison believed to be just as effective as the brass behemoths he had been building for three decades.

Harrison's story, including how he had to petition George III directly for payment of the prize money he believed he was owed, is superbly told by Dava Sobel in her book *The Illustrated Longitude*, and cannot be improved upon by this author. In Harrison, Sobel saw a self-taught maverick, battling the goliath of a lofty and unsympathetic establishment. For much of the eighteenth century, the scientific consensus was that the reliable way of determining longitude was not through man-made timepieces, but by using the heavens. This 'establishment' opinion was perhaps best articulated by the Westminster- and Cambridge-educated fifth Astronomer Royal, the Reverend Nevil Maskelyne. And, as Sobel puts it, 'A story that hails a hero must also hiss at a villain – in this case, the Reverend Nevil Maskelyne, remembered by history as the "seaman's astronomer".' [9]

Maskelyne was an eighteenth-century high flyer: fellow of Trinity College Cambridge and by his early twenties a protégé of the third Astronomer Royal, James Bradley, he was elected to the Royal Society aged twenty-five. In 1761, they sent him to St Helena to observe the transit of Venus – clearly the expert on the annual parallax of Sirius was just the man for the job. In his early thirties, he became Astronomer Royal. It was an astonishing ascent and he could rise no further.

On his way down to the island that Napoleon would later call home, he had experimented with a system of determining the longitude by lunar observation. Maskelyne believed that he was on the brink of solving the longitude problem with the lunar-distance method, which used a system of celestial landmarks. A set of tables informed mariners of the times that the moon could be observed to reach certain stars over a given location. By taking their local time and having made the relevant astronomical observation, they could ascertain longitude by calculation. Of course, this required a cloudless night, calm conditions under which to make the observation, as well as detailed knowledge of the heavens and around four hours in which to make the necessary observations and calculations. As an astronomer, Maskelyne favoured this method. As Astronomer Royal, he had a seat on the Board of Longitude, and with every year that Harrison was taking to perfect his horological solution, knowledge of the heavens was growing, and the lunar table method was looking ever more credible. Harrison began to fear that he might be overtaken by a precocious young man who had been just three years old when he had presented H1 to the Board.

By now in his late seventies, ailing, and despairing of ever being awarded the prize, Harrison applied directly to the king, now George III. The monarch tested Harrison's latest timepiece, H5, in his own private observatory at Kew, and, with the words 'By God! Harrison, I will see you righted!', pressed Parliament to pay Harrison a sum that was almost equivalent to the balance of the longitude prize. Further vindication followed when Harrison allowed a copy of H4 to be made. It accompanied Captain James Cook on his epic three-year voyage of discovery that included Antarctica and the tropics. Its performance was exemplary. On 24 March 1776, just eight months after Captain Cook's return, Harrison died aged eighty-three, but he went to his grave knowing that the object to which he had consecrated the greater part of his life had been achieved. It was now left to a younger generation of watchmakers, most notably John Arnold, Thomas Earnshaw and Thomas Mudge, to develop series production of marine chronometers.

But that does not mean that history proved Maskelyne wrong. By the mid-1760s, Maskelyne had succeeded in reducing the time needed for calculation to just half an hour when using his *Nautical Almanac and Astronomical Ephemeris*. Besides, according to Cambridge academic Dr Alexi Baker, whose detailed four-part rehabilitation of Maskelyne appeared on the National Maritime Museum website in 2011 (the bicentenary of the astronomer's death), marine chronometers of the type made by Harrison 'were not cheap enough to be widely used until the 1800s'. After all, if the relevant timekeeping equipment cost about a third as much as a sailing vessel, the practical application of Harrison's solution would have been limited. 'In the meantime, the pursuit of an accelerated lunar-distance method led to other improvements in astronomy and navigation, the establishment of the Nautical Almanac which is still being published, and the development of the basic sextant design that is still in use today.' [10]

It is hard to say whether history favours the autodidact Harrison's impossibly accurate marine chronometers or the tidy-minded Maskelyne with his tables, maps and charts. But the advances born of the intellectual duel between the two men contributed significantly to the maritime supremacy that enabled Britain to emerge from the eighteenth century as the pre-eminent colonial power and, ultimately, possessor of the greatest empire the world has yet seen.

James Cook (1728–1779), English navigator, explorer and hyrdrographer, took an Arnold chronometer with him on his second exploratory voyage to the South Seas in 1772. He is seen here receiving the tributes of the Sandwich Islanders. Not all indigenous peoples were as friendly: the inhabitants of Hawaii were less well-disposed towards him and killed him, thus bringing his voyages of exploration to an abrupt end.

1. Dava Sobel and William J. H. Andrewes, *The Illustrated Longitude* (New York City: Walker & Co., 2003), p. 17
2. Derek Howse, *Greenwich Time and the Discovery of the Longitude* (Oxford: Oxford University Press, 1980), p. 47
3. *House of Commons Journal*, 25 May 1714, quoted in Derek Howse, *Ibid.*
4. Warrant for foundation of Royal Observatory, quoted in Derek Howse, pp. 50–51
5. *Ibid.*
6. *Ibid.*
7. Bank of England inflation calculator (https://www.bankofengland.co.uk/monetary-policy/inflation/inflation-calculator)
8. Dava Sobel and William J. H. Andrewes, *The Illustrated Longitude* (New York City: Walker & Co., 2003), p. 93
9. *Ibid.*, p. 138
10. 'Rehabilitating Nevil Maskelyne Part Four: The Harrisons' accusations, and conclusions', Royal Museums Greenwich blog, 12 February 2011 (https://www.rmg.co.uk/discover/behind-the-scenes/blog/rehabilitating-nevil-maskelyne-part-four-harrisons-accusations-and)

Time With *a* Bang
Canon Solaire

Louis-Philippe Joseph d'Orléans: nobleman, revolutionary and the visionary behind the transformation of the Palais Royal. A political radical, after the Revolution he was named Philippe Égalité and he voted for the death of his relative King Louis XVI. His political credentials notwithstanding he was subsequently guillotined himself. He was the father of France's last monarch, King Louis Philippe.

During the second half of the 1780s, the Duke of Orléans was one of the richest men in the world. In 1787, his rent roll amounted to 7,500,000 livres (over £48 million in today's money) and his sprawling estates 'ran to the extent of three or four of today's départements'.[1]

As the head of the cadet branch of the French royal family, these great riches came with great power. His great-grandfather had been Regent of France during the Minority of Louis XV, and he would father a future king.

In the Paris of the 1780s, this concentration of wealth and power found its physical expression in the Palais-Royal – an architectural metaphor for the progressive, fashionable young Anglophile duke who had succeeded his father in 1785. In the words of historian George Armstrong Kelly, 'his Palais-Royal dominated Paris as Versailles dominated France'.[2]

Originally built by Cardinal Richelieu during the 1630s, on his death it had become Crown property and changed its name from Palais-Cardinal to Palais-Royal. When Louis XIV's younger brother Philippe, Duke of Orléans, married the daughter of Charles I of England, the Palais-Royal became the principal residence of the House of Orléans, and the centre of fashionable life in Paris. In 1780, the then Duke of

A view of the elegant gardens of the Palais Royal by day.

Vue perspective du Palais Royal du coté du Jardin

A typical evening at the Palais-Royal.

Orléans gave the Palais-Royal to his son, who set about an ambitious remodelling, which included developing the gardens to the rear, enclosing them with three arcaded rows of shops, and adding upper storeys of apartments for rent.

Novel and controversial, it was to be a combination of gated residential community, urban resort and shopping mall. The Palais-Royal is often credited with changing the way people shopped, and making retail history, opening the age of the arcade that endured until the 1930s. As well as shops, cafés and a theatre, the central space was modelled on the pleasure gardens of Vauxhall and Ranelagh. As Fragonard captured the pleasure-seeking *douceur de vivre* of this period on canvas, so the Duke of Orléans and his architect Victor Louis explored the same spirit of elegant, elaborate frivolity in architecture.

Paris had never seen anything like it.

Unlike London, the streets of Paris were unpaved. Treading them was a dirty and dangerous business; even if one avoided being run over by the carelessly driven carriages of the rich, it was hard not to be splashed with water, mud and worse in the narrow, foetid streets and alleyways of pre-Hausmann Paris. At night, the unlit streets were even more hazardous.

By contrast, the arcades of the Palais-Royal were lit, paved and protected by a private police force. And, unlike the rest of Paris, the rising sun saw an army of cleaners at work preparing the Palais for another day's dissipation.

This early nineteenth-century dial, made by Rousseau, consists of a round marble slab engraved with a sundial. A small brass cannon holding a tilting lens is attached to the slab. At noon, the sun's rays, focused by the lens, ignite the gunpowder and thus cause the cannon to fire. For this reason the instrument is called a noon cannon.

Visitors were stunned.

'One could spend an entire life,' wrote one, 'even the longest, in the Palais-Royal and, as in an enchanting dream, dying say "I have seen and known all".' [3]

Night was particularly bewitching.

> *'Although the arcades shed their light upon the green boughs, it was lost among the shadows. From another walk floated soft, sweet sounds of tender music. A slight breeze stirred the tiny leaves of the trees. "Nymphs of joy" approach us one after the other, threw flowers at us, sighed, laughed, invited us into their grottoes with promises of untold delights, and vanished, like phantoms of a moonlit night.'*

It was the best of the city in one place: 'Everything that can be found in Paris (and what cannot be found in Paris?) is in the Palais-Royal.' [3] And among the many things to be found in this palace of pleasures was the correct time.

Between the months of May and October, some time before noon, a fashionable crowd would gather in the gardens of the Palais-Royal, their watches in hand. The fewer clouds in the sky, the greater the crowd in the gardens. At solar noon, as the sun passed the meridian, the expectant crowd was rewarded with a small explosion.

In 1786, an ingenious, marketing-minded watchmaker called Rousseau who had a shop in one of the arcades, at No. 96 Galerie de Beaujolais, had placed a small bronze cannon on a plinth in the middle of the gardens. It was equipped with a carefully positioned powerful magnifying lens that concentrated the sun's rays as it reached solar noon, so that they ignited a wick and detonated the charge. As the

The ritual detonation of the cannon at solar noon attracted men about town and transformed the hitherto banal action of setting one's watch into a fashion statement.

report of this solar-powered piece of artillery echoed around the elegant arcades, the men of fashion synchronised their watches.

It was an integral part of what today would doubtless be called 'the Palais-Royal Experience', and was captured by the poet Jacques Delille, who, walking through the Palais-Royal with the Duke of Orléans one day, was asked to write a few lines of epigrammatic verse expressing his opinion of the place. He obliged with a witty quatrain:

Dans ce jardin tout se rencontre,	In this garden you find everything,
Excepté l'ombrage et les fleurs;	Except shade and flowers;
Si l'on y dérègle ses moeurs,	If your morals go wrong there,
Du moins on y règle sa montre.	At least you can set your watch right.

However, Paris Society soon had more to worry about than setting its watches. The French Revolution overshadowed this world of artificial gaiety and, ironically, given his background, the Duke of Orléans embraced these changes, called himself Philippe Egalité, and the Palais-Royal, renamed Palais-Egalité, thronged with *sans-culottes* as well as pleasure seekers. He even voted for the death of his relative King Louis XVI, yet he was later guillotined himself. His son escaped and lived to become King Louis Philippe, the last French monarch.

The solar cannon sat out the Revolution in one of the Palais-Royal's cafés and, restored to its plinth in the gardens, resumed its duties in 1799. However, after 1911, when Greenwich meantime was adopted, it fell into disuse. For a while it was protected by a glass case, its distinctive appearance suggesting to one writer 'a bronze toad in its little aquarium'.[4] By the early 1970s, however, it was in a state of decay: oxidized, missing its magnifying glass and much of the attendant assembly.

In 1975, a restoration project was undertaken and in May a functioning cannon was returned to the Palais-Royal. As well as the simple solar noon, the new cannon could be adjusted to take account of the difference in time between the meridians of Paris and Greenwich; the equation of time; and the seasonal changing of the clocks.

It was inaugurated on 14 May 1975, when the sun would cross the meridian at 12.47 p.m. Alas, at 12.45 a passing cloud spoiled the spectacle. The cannon was stolen in 1998 and a simplified, non-functioning replica put in its place in 2002.

These days the cannon, rather like the Palais-Royal, is quiet and peaceful. The setting of watches is no longer the major social event it once was. Moreover, those crowds that gathered during the 1780s were not, strictly speaking, setting their watches to true Paris time. The cannon was not positioned precisely on the meridian of Paris, which was sixty metres or so to the west, but a deviation of such small distance was an acceptable sacrifice for fashion.

In his 1979 essay on the solar cannon of the Palais-Royal, Louis Marquet makes the ingenious point that 'the speed of sound being more or less equal to the "movement" of the sun at that latitude, people to the west of the cannon, in the direction of the Arc de Triomphe, hearing the detonation set their watches to the true midday of the place where they found themselves.'[5]

Such sophistry aside, the clock-watchers of the Palais-Royal would have doubtless felt it was far more important to be *en vogue* than on time.

---

1. George Armstrong Kelly, 'The Machine of the Duc D'Orléans and the New Politics', *The Journal of Modern History*, Vol. 51, No. 4 (1979), pp. 667–84
2. *Ibid.*
3. Quoted in *Country Life*, 30 January 1986
4. Louis Marquet, 'Le canon solaire du Palais-Royal à Paris', *L'Astronomie*, Vol. 93 (1979), p. 369
5. *Ibid.*

American Polymath
Franklin

Politician, postmaster, publisher, scientist, satirist, diplomat, optician, oceanographer, moralist, meteorologist, monetary theorist, revolutionary, demographer, kite flyer, chess player, composer, educator, statesman, polyglot… Benjamin Franklin was less of a regular human being, more a human Swiss Army knife. If nothing else, he was a very useful man to have around the place – whether you wanted a peace treaty negotiated (the Treaty of Paris 1783), your eyesight needed improving (bifocals were his idea), you had a nasty urinary complaint (he designed a flexible silver catheter for his brother John who suffered from kidney stones) or simply wanted to shrug off the oppressive yoke of tyranny and send one of the world's great military powers packing (the small matter of the American War of Independence). It seems there is nothing that Franklin could not turn his hand to, so it is no surprise that he was also a talented horologist.

Gallery 553 may not be the busiest of the Metropolitan Museum's exhibition spaces; the hundreds of thousands who throng the annual Costume Institute Exhibitions tend to give this small room of neoclassical furniture and objects a miss, unless they happen to pass by it on the way to get a snack at the nearby Petrie Court cafeteria. But those who do stop here will be familiar with a strange-looking clock housed in an obelisk-shaped case of oak and thuya burl wood.

One of the more unusual timepieces in the Metropolitan Museum of New York is this late eighteenth-century 'Obelisk Clock' with a Franklin movement. As well as accommodating the swing of the pendulum, the Egyptian-style obelisk shape of the case, along with the finial the double-faced head of Janus, reflect a rising interest in the civilizations of Classical Antiquity.

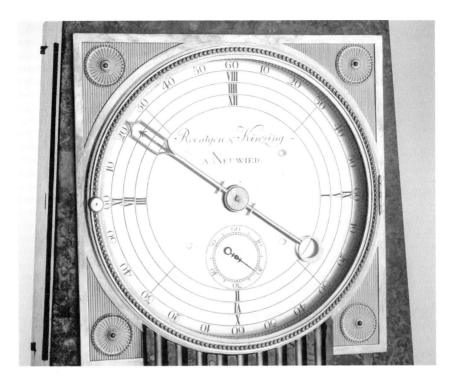

Benjamin Franklin intended for his simplified movement – consisting of just three wheels, a four-hour spiral-ring dial and a single minute hand – to be used in economically produced clocks. However, in this instance, it has been used to make a luxurious timepiece created by two of Marie Antoinette and Louis XVI's most favoured craftsmen, cabinetmaker David Roentgen and clockmaker Peter Kinzing, with whom Franklin may have come in contact while in Paris.

The dial bears the names of David Roentgen and Peter Kinzing. The two men were favourites of King Louis XVI and Marie Antoinette, whom they supplied with extravagant and elaborate mechanical objects. Roentgen even bore the title '*ebeniste mecanicien du Roi et de la Reine*' – 'cabinet maker of the King and Queen'. Typical of the inventive duo's work was a mechanical desk with numerous secret drawers, featuring a clock made by Kinzing that played a dozen different melodies, for which the king (a noted lover of clocks) parted with an impressive 80,000 livres (the equivalent of over £500,000 today).[1]

However, it is not the names on the dial that make this clock interesting, but the dial itself. It features just a central minute hand, no hour hand, and four concentric scales: the outer is marked in four quadrants calibrated for sixty minutes each and each of the three subsequent circular scales is calibrated for four hours. The hand makes one circuit of the dial every four hours and the time in hours can be read spirally from one to twelve. It was invented by neither Roentgen nor Kinzing, but, of course, Benjamin Franklin.

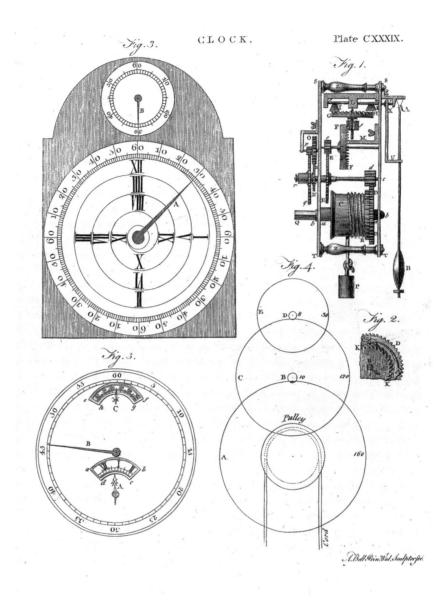

The movement of Franklin's clock was the embodiment of simplicity as this engraving shows. Engraving by Andrew Bell (1726–1809).

It is, as most of his ideas were, sheer genius, drastically reducing the number of components required to make a timepiece.

The invention is thought to date from 1758, when it first appeared by the name of 'Ferguson's Clock', after the British clockmaker who perfected Franklin's design for a weight-driven clock with a movement comprising just three wheels and two pinions.

Franklin the polymath, surrounded by symbols of his wide-ranging genius. Sadly this image does not appear to allude to his prowess as a comic writer.

It is a masterpiece of economy and simplicity; moreover it has an ideological, almost moral, dimension in that it makes the time available with a minimum of components. 'Several clocks have been made according to this ingenious plan of the Doctor's [Franklin, of course, held doctorates galore], and I can affirm, that I have seen one of them, which measures time exceedingly well,' noted Ferguson approvingly. 'The simpler that any machine is, the better it will be allowed to be, by every man of science.'[1]

There is of course an amusing irony about the clock in the Met: the Puritan-born democrat and Republican's super simple movement has been used – one might even say hijacked – by the eighteenth century's foremost creators of luxury mechanical items for Europe's most decadent Royal Court.

But then, as a noted early American humourist – yes, apparently some of his early writing was hilarious[2] – Benjamin Franklin would have no doubt enjoyed the joke.

---

1. James Ferguson, 'Account of Franklin's Three-Wheel Clock, 1758', *Founders Online*, National Archives, last modified 13 June 2018 [http://founders.archives.gov/documents/ Franklin/01-08-02-0060] [Original source: Leonard W. Labaree (ed), *The Papers of Benjamin Franklin, Vol. 8, April 1, 1758, through December 31, 1759* (New Haven and London: Yale University Press, 1965), pp. 216–20]

2. Benjamin Franklin, *The Sayings of Poor Richard: Wit, Wisdom, and Humor of Benjamin Franklin in the Proverbs and Maxims of Poor Richard's Almanacks for 1733 to 1758*

Guillotines *and* Grand Complications
Marie Antoinette's Breguet

𝒫arking his small Simca 1000 hatchback, the hollow-cheeked, sharp-featured man turned the engine off and waited for a moment, his intelligent eyes taking in his surroundings. Getting out of the car, he went to the boot, lifted out a tool box and strolled towards a large building of honey-coloured stone.

Working quickly and quietly, he used a car jack to part the bars of a gate and then wriggled through.

It was the evening of Friday 15 April 1983; the building of honey-coloured stone was Jerusalem's L.A. Mayer Memorial Institute for Islamic Art; and the biggest burglary in the history of Israel had just begun.

The museum had been opened nine years earlier. Founded by the late Vera Salomons and named in honour of her friend, scholar and archaeologist Professor Leo Ari Mayer, it houses one of the world's most significant collections of Islamic Art.

But, as well as the jewellery, glassware, carpets and ancient pages of the Qur'an, the museum is home to a collection of unique timepieces, many made in France during the late eighteenth and early nineteenth century by the horological genius often regarded as the finest watchmaker ever. One in particular, known dramatically, if not inaccurately, as 'the Mona Lisa of timepieces', was the most

The 'Marie Antoinette', arguably the most famous watch in the world.

celebrated watch in the world. Its importance, its value and its complexity were belied by the name, or, rather, the number assigned by its maker: 160.

The 160th entry in the pages of an eighteenth-century watchmaker's order book, this watch is better known by the name of its intended recipient: Marie Antoinette.

To understand what a watch made for the most famous queen of France was doing in a museum of Islamic art in Jerusalem, it is necessary to travel back to 1762, to the lakeside town of Neuchâtel in what today is modern Switzerland, where a 15-year-old boy was boarding the stagecoach to Paris. His father had recently died, and his mother had remarried. Her new husband was a watchmaker and, after a year as his apprentice, her son demonstrated such gifts that he headed for Paris, where he would make his name, his fortune and history.

There are some individuals so important to human development that they mark a watershed: Copernicus and Galileo did with astronomy, Columbus with exploration, Shakespeare with the English language, Picasso with painting and Abraham Louis Breguet, the teenager on the stagecoach travelling the rutted roads towards Paris with the watch. Breguet changed the personal timepiece more profoundly than any single individual before or since. In short, he either invented or improved most parts of the mechanical watch as we know it today.

Even the most cursory survey of Breguet's main achievements is impressive: in 1780, he brought out his first self-winding watches; three years later, he invented the gong spring for repeater watches; in 1790, he came up with the pare-chute shock-absorption system; in 1796, he brought out the first carriage clock; and, of course, he is best known for his 1801 patent, the tourbillon. Within watchmaking, he is remembered countless times on a daily basis in the adjectival use of his name to describe various aesthetic and technical aspects of the craft: there are Breguet hands, Breguet numerals and the Breguet overcoil. More than a gifted horologist, he had a taste for flamboyant marketing. Once, in order to show off his new shock-absorption system, he took out his watch in front of Talleyrand and threw it on the ground.

Needless to say, he was the darling of French Court circles.

However, Breguet also liked to live dangerously, and when the seething discontent of the French people exploded into the orgy of bloodletting that was the French Revolution, he was caught up in the excitement of the liberation of the Bastille and the publication of the *Declaration of the Rights of Man and of the Citizen*. He joined the Society of the Friends of the Constitution, which soon became known by the name of the former Jacobin monastery, near the Tuileries, where its members gathered.

Only the massacres of September 1792 finally tempered Breguet's radicalism and, becoming more moderate, he fell foul of Robespierre. Fearing for his life, he asked his old friend Jean-Paul Marat, who had also come to Paris

Abraham-Louis Breguet changed the mechanism of the personal timepiece more profoundly than any single individual before or since.

from Neuchâtel, to help him leave the country. On 24 June 1793, the National Convention Committee for General Security and Surveillance heard the case of Citizen Breguet and agreed to grant him and his immediate family a passport. It was just in time. A few days later, on 13 July, Marat was killed in his bath.

Breguet spent some tense weeks waiting for his papers, but they came through and on 11 August he left the business he had spent thirty years building. Among the things he packed and took with him into exile was the unfinished project referred to as 'No. 160'. Not knowing what would happen to him or whether he would ever return to Paris, he must have looked at the assemblage of components with mixed emotions.

The order had entered his heavy ledgers exactly a decade earlier under very mysterious conditions: placed by an unknown officer of the Queen's Guard.

Whether Her Majesty was aware of the order and on whose behalf the officer was acting are unknown. This mystery is all the more frustrating because what Breguet had been asked to design was the definitive portable timepiece, encyclopaedic in its scope, incorporating the full range of functions known at the time: 'a minute repeating *perpetuelle* watch, with complete perpetual calendar, equation of time, power-reserve indicator, metallic thermometer, large optionally independent seconds-hand and small sweep seconds-hand, lever escapement, gold Breguet overcoil, double *parechute,* all points of friction, holes and rollers, without exception, in sapphire, gold case, rock crystal dial and gold and steel hands.'[1]

Pages of a production ledger recording the stages of manufacture of Breguet no. 160, the grand complication – the watch known as the 'Marie-Antoinette'. (From the Collection Montres Breguet.)

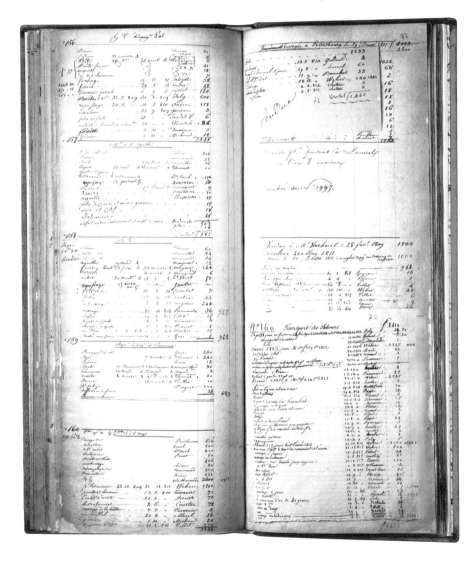

Forerunner of today's self-winding watches, the 'Perpetuelle' watch features a sprung weight on a pivoting arm that jumps up and down in response to the wearer's movements when walking, to wind the mainspring.

The order included neither deadline nor maximum price; the only stipulation was for gold to be used wherever possible. In effect, Breguet was being asked to produce a cathedral clock of the sort that had been the most technically advanced objects of the Renaissance – only this time within the confines of a pocket watch. It was the ultimate horological high-wire act – a feat of unparalleled technical sophistication and mechanical miniaturisation destined to be worn and admired at the leading royal court in Europe.

It was not Marie Antionette's first Breguet: in 1782, he made her the 'perpétuelle', a self-winding repeater watch fitted with a calendar. Her husband the king, known to be fascinated by mechanical objects and watchmaking, was also a Breguet customer. However, the man behind the commission of No. 160 was not her husband, but widely suspected to have been a Swedish count, Axel von Fersen, rumoured to be the queen's lover. An alternative theory is that 'the watch was intended as a present to one of the Queen's favourites'[2] – perhaps Fersen.

Whatever the truth, it was a gift she would neither give nor receive. Only a matter of weeks after Breguet fled the French capital she went to her death at the guillotine, in front of the baying crowd in the blood-soaked Place de la Revolution (now the more peacefully named Place de la Concorde).

Breguet was luckier than his client. He would return to Paris and enjoy even greater glory as the watchmaker by appointment to the Napoleonic elite. Indeed, one of the earliest recorded wristwatches was made by him for Napoleon's sister Caroline Murat, Queen of Naples. It was even speculated that the French emperor 'frequently went incognito to the workshop and conversed upon the improvements which he

Visitors admire Breguet watches, part of the exhibition 'Mystery of Lost Time' at the Museum for Islamic Art in Jerusalem, July 21, 2009, at which the 'Marie Antoinette' was displayed for the first time since its theft.

was anxious to effect in cannon and fire-arms'.[3] In all, Napoleon's family accounted for around 100 pieces between 1797 and the derailment of the Bonaparte gravy train in 1814. But regime change just brought Breguet more customers, not least the Emperor Alexander of Russia and the Duke of Wellington; the latter is said to have paid 300 guineas for a repeater (around thirty times what a cavalry man in the Scots Greys was paid in one year).

Breguet's extraordinary career spanned Bourbon, Revolutionary, Napoleonic and Restoration France, but, throughout this sprawling historical novel of a life, one thing remained constant: even in his final days, he was still working on No. 160. It was not completed until 1827 – thirty-four years after Marie Antoinette's demise at the

guillotine; seventeen years after Fersen's
life ended at the hands of a lynch mob;
and four years after Breguet himself
died (he had left instructions in his will
for his son to complete the work).

Count Hans Axel von Fersen.

It was to retain the undisputed
title of the world's most
complicated watch for a
century, and its subsequent
history was just as
fascinating as the events
surrounding its creation.
Purchased by a Marquis
de la Groye, who as
a child had been one
of Marie Antoinette's
pages, it was returned for
repair in 1838, but never
collected, and remained
with Breguet in Paris until
1887, when it was sold to a British
visitor, Sir Spencer Brunton. It passed
to his brother and was then bought by
collector and art dealer Murray Marks, whose
clients included
J. P. Morgan. He sold it to Louis Desoutter, a famous restorer of Breguet watches,
and then, one rainy spring day in 1917, in the window of a shop in London's West
End, it caught the eye of Sir David Lionel Salomons: it was a *coup de foudre*.

> *My attention was attracted by a curious-looking watch differing from the usual display, and*
> *I saw a notice by its side, bearing the name "Marie Antoinette." I then went up to the window*
> *to have a better look at the watch, and I saw that it had been made for that ill-fated Queen by*
> *Breguet, and was his masterpiece. A high price was put on it.' But it was worth it. 'It turned out*
> *to be a good purchase, judging from seducing offers made to me later on to part with it. Evening*
> *after evening, I studied this watch, which is most complex and interesting, with the result that*
> *I formed the opinion that no other maker of watches could approach such work.*[4]

On Salomons' death, it passed to his daughter Vera, who died in 1969. Five years
later it went on public display in the Museum of Islamic Art when the building
was completed and opened in 1974. There it remained until that warm spring
evening in 1983.

The same year as the enterprising thief squirmed through the gate and windows to steal this piece of horological history, the future of watchmaking was launched in Switzerland.

Where No. 160 had been the costliest, the Swatch played on its accessible price; where No. 160 had taken forty years to make, Swatch was designed to be made swiftly and easily; where there was just one No. 160, now vanished, Swatch could, and would, be churned out in millions. This plastic watch would become one of the most familiar objects on the planet. Indeed, so successful would it become that in 1999 Swatch would buy the storied old firm of Breguet.

The architect of Swatch's success was a Lebanese-born management consultant called Nicolas G. Hayek, a man often described as the saviour of the Swiss watch industry. In later life he became so obsessed by Breguet that he was jocularly known in the industry as Abraham Louis Hayek.

Frustrated that, twenty-one years after the robbery, No. 160 remained lost to humanity, he put the formidable resources of the Swatch Group behind recreating the lost treasure. Even with the technology of the twenty-first century, the replica would take four years to build. But, in 2006, before it could be unveiled, the dramatic deathbed confession of the thief, Na'aman Diller, a former military pilot turned forger, burglar and criminal polymath, revealed that after the watches had been stolen, they were wrapped in newspaper, placed in boxes and consigned to a warehouse, where they had remained for twenty-one years. Among them was No. 160.

It has now been restored to the museum where, protected by rather more exigent security measures, it waits to see whether the coming centuries will be as eventful as its first two.

---

1. Emmanuel Breguet, *Breguet: Watchmakers Since 1775* (Paris: Gourcuff, 1997), p. 48
2. Sir David Lionel Salomons, *Breguet* (London: 1921)
3. Rees Howell Gronow, *Captain Gronow's Last Recollections: Being The Fourth And Final Series Of His Reminiscences And Anecdotes* (Palala Press, 2015), p. 76
4. Sir David Lionel Salomons, *Breguet* (London: 1921), p. 4

Time Delivered *to* Your Door
The Belville Chronometer

'*Miss Elizabeth Ruth Belville, who has died at Wallington, Surrey, at the age of 89, devoted half a century to taking the correct Greenwich time to business houses in London on a watch 100 years old. Three times a week she went to Greenwich, where she obtained a certificate of accuracy for her watch.*' [1]

The Times greeted its readers on the morning of 13 December 1943 with the familiar front page crowded with advertisements, births, deaths, personal announcements, legal notices and public appointments: a gentleman's musquash coat, 'perfect condition', was advertised for sale in the West End; Beaconsfield school required

A Victorian couple looking at the Shepherd Gate Clock on the wall outside the gate of the Royal Observatory at Greenwich. In those days Greenwich Meantime was a valuable commodity to be sold in the West End.

*Above: How Miss Belville appeared to readers
of* Popular Science Monthly *in 1929 as
The Clock-Woman of London.*

*Left: The clock at the Royal Observatory,
Greenwich, more recently.*

an assistant master – 'Latin and French essential'; a 'gentleman' announced that he
wanted to buy 300 or 400 Havana cigars; and Leicester Royal Infirmary was looking
for a masseuse. Inside was equally packed with world events: the battle on the Kiev
salient; the Czech–Soviet treaty; U-boats in the Atlantic; the future of the monarchy
in Greece; the announcement in Algeria that General de Gaulle was giving tens of
thousands of Muslims French citizenship…

All in all, the reader of 'The Thunderer' that day could have been forgiven for
missing the fifty-five words squeezed in just above the crossword at the bottom of
page six, next to a slightly longer obituary of the Croydon Corporation catering
manager. However, with the death of Miss Belville, a little bit of British timekeeping
history had disappeared.

Her grandmother had arrived in Britain as a refugee from the French Revolution
and had given birth to a son, John Henry, in the summer of 1795.[2] When the boy
was about five, she died and he was adopted by John Pond (it is suggested that Pond
was his father).[3] Pond succeeded Harrison's rival in the search for longitude, Nevil

Maskelyne, as sixth Astronomer Royal, and in 1816 John Henry joined him as second assistant. He would continue to work at the Royal Observatory until his death forty years later.

By the 1830s, it was clear that Pond was suffering from a debilitating mental illness and that he had let the Observatory fall into disarray. He was replaced by the zealous Professor George Airy, who lost little time in establishing some order. Chronometer makers had been in the habit of sending messengers to the Royal Observatory to get the correct time, but Airy found these visits intrusive, restricted them to Mondays, and agreed to the idea – mooted by his predecessor – that John Henry Belville be placed in charge of distributing the time. In the twenty-first century, accurate time is ubiquitous and has been easily available to anyone with a radio or a telephone for generations. In the 1830s, however, the right time straight from the 'manufacturer' in Greenwich was a prized commodity.

Maria Belville (1811–1899), the third wife of John Henry Belville (1794–1856): on his death she continued his duties 'delivering' Greenwich Mean Time to the West End.

Before his illness had incapacitated him, Pond had installed the time ball on the roof of the Observatory, which dropped at 1 p.m. every day (1 o'clock was chosen to allow astronomers sufficient time to make the necessary observations to determine noon exactly). It could be seen by ships moored on the Thames, enabling them to regulate their chronometers.

Airy went further.

Belville was put in charge of a newly installed twenty-four-hour clock at the Observatory gate, and in June 1836 [4] he started making a weekly trip to the West End and City to give the city's chronometer makers, and others who paid an annual subscription for the service, Greenwich time, as transported by his watch direct from the Observatory.

Like fish caught that morning or bread warm from the baker's oven, time was best enjoyed when served as fresh as possible. With the opening of the capital's first steam railway between London and Greenwich in 1836, which conveyed passengers above the market gardens and growing suburbs of south London on a viaduct of 878 brick arches, Greenwich time could be served piping hot in central London within minutes.

By the time John Henry Belville died in 1856, time signals conveyed by telegraph and galvanism (as electricity was known at the time) were already available, but according to Professor David Rooney, author of a wonderful short book on the subject, 'many of John's subscribers were keen to continue with the technology they knew and trusted'.[5]

Maria was Belville's third wife, and the couple had recently had a daughter, Ruth, who was two years old at the time of his death. Maria wrote to Airy requesting a pension from the Admiralty. When that appeal was rejected, she sought and received permission to continue her husband's work, bringing her daughter with her on her rounds.

Business was so brisk that there was even a secondary market for the time sold by the Belvilles, as Ruth recalled later in life:

> I myself have a sort of recollection of a firm in Clerkenwell where I went with my mother when I was a small child… after she had checked the regulators by the chronometer… we passed three or four people going in to the shop, chronometer in hand, and my mother telling me that these people were working chronometer-makers who paid a small fee to the big firm so that they might obtain the time second hand![6]

Moreover, by the time Maria Belville retired in 1892, there was still enough demand for her daughter to take over the business.

Known internationally as 'The Clock-Woman of London', photographs taken of Ruth Belville show a woman who comes across as a mixture of Miss Marple and Mary Poppins, ever capable and ever reliable.

The Belville family's pocket watch, 'Arnold'. By the time this watch finished its service, it had carried Greenwich time to the heart of London for over a century.

Bustling about the capital, Belville became something of a celebrity. A walking London landmark, she was interviewed in the press and appeared more than once in the pages of *Popular Scientist* magazine, where she was known to readers by the affectionate sobriquet 'The Clock-Woman of London'.[7] Photographs taken of her in her seventies show a woman who comes across as a mixture of Miss Marple and an elderly Mary Poppins, ever capable and ever reliable, in a neat little hat and old-fashioned ankle-length coat, giving the time to her clients with a brisk authority, be he a gasman or an office worker standing on the tips of his toes on a stepladder adjusting the hands of a large wall clock as she gives instructions from below, chronometer in hand.

The source of her fame and her livelihood, the large pocket chronometer was the same one used by her mother and before that her father when he commenced the service. Given that she finally stopped her work in 1940, when the war made the streets of London dangerous (she was over eighty-five at the time), by the time this watch finished its service, it had carried Greenwich time to the heart of London for over a century.

Astonishingly, the watch was older even than the service it provided. It had been made before John Henry's birth, in 1794, by the chronometer-maker John Arnold, and was still being carried around London in Miss Belville's handbag at the outbreak of the Second World War.

Understandably, she was very attached to this venerable timepiece, and over the years she developed a somewhat anthropomorphic relationship with it: 'She always referred to the watch as Arnold, as if it were the Christian name of a dear friend,' recalled one who knew her in later life. 'Her business with a client would be performed something like this: "Good morning, Miss Belville, how's Arnold today?"

– "Good morning! Arnold's four seconds fast today," and she would take Arnold from her handbag and hand it to you… The regulator or standard clock would be checked and the watch handed back. That would be the end of the transaction for a week.' [8]

A historically significant watchmaker, John Arnold had known both Breguet and Harrison (of longitude fame). Using the principles established by the latter, he had done much to advance marine chronometry. He also made watches for the royal family; indeed, the Belville watch had been intended for the Duke of Sussex, one of George III's sons, but he had rejected it on the grounds that it was too big.

By no means the grandest of Arnold's watches, the Belville chronometer is now exhibited in the Science Museum in London. An unassuming-looking, silver-cased timepiece, the eye easily drifts across it. It could be argued that this is the watch that most fully expresses Arnold's genius as a watchmaker, when it is considered that a timepiece he had made in the days of Pitt, when men wore wigs and breeches, continued to be used in an exacting professional capacity until the time of Winston Churchill and the threshold of the atomic age.

1. *The Times*, 13 December 1943, p. 6
2. J. L. Hunt, 'The Handlers of Time: The Belville Family and the Royal Observatory, 1811–1939', *Astronomy & Geophysics*, Vol. 40, Issue 1 (1 February 1999)
3. *Ibid.*
4. *Ibid.*
5. David Rooney, 'Ruth Belville: The Greenwich Time Lady', Science Museum Blog, 23 October 2015
6. Quoted in David Rooney, 'Ruth Belville: The Greenwich Time Lady', Science Museum Blog, 23 October 2015, p. 52
7. *Popular Scientist*, October 1929, p. 63
8. Quoted in Derek Howse, *Greenwich Time and the Discovery of the Longitude* (Oxford: Oxford University Press, 1980), p. 87

The Most Famous Clock *in the* World
Big Ben

For hundreds of years the Exchequer had calculated taxation with wooden tally sticks. However, in 1834, two years after the Great Reform Bill, the clerk of works at the Palace of Westminster came to the unsurprising conclusion that these memory aid devices were old-fashioned, out of step with modern times and had to be disposed of.

The eradication of pre-Reform paraphernalia was to prove more thorough than he, or anyone, could have envisaged.

Instructions were given that the wooden counters be used for firewood. But, instead of giving them to the poor, workmen stuffed them into two furnaces under the House of Lords. With Parliament in recess and the Palace of Westminster left in the care of a housekeeper, Mrs Wright, no one really noticed the intense heat building up in the centuries-old chimney flues of the medieval building. By 4 o'clock on the

Somewhat more prosaic than Turner's interpretation of the conflagration, this is the view of the destruction of both Houses of Parliament as seen from the Thames.

The Great Clock of Westminster in all its Gothic Revival splendour.

afternoon of 16 October, however, two visitors to whom Mrs Wright was showing the House of Lords were barely able to make out the tapestry-covered walls, so thick was the smoke, while beneath their feet the stone floor radiated a heat that could be felt through their shoes. Even though they could not descry much detail, they had the distinction of being the last people on Earth to the see the inside of the old House of Lords.

The Palace of Westminster's famous clock tower against a backdrop of New Year's Eve fireworks.

Mrs Wright locked the door of the chamber at 5 p.m. At about 6 o'clock the flickering of flames was noticed at the bottom of the door of the House of Lords. Minutes afterwards, the building was ablaze.

The British Houses of Parliament were on fire and burning brightly, lighting up the autumn evening. Fire engines and firemen were sent to the blaze, as were soldiers and a detachment of Robert Peel's new-fangled police force. But most of the venerable buildings from which Britain had been governed for centuries were beyond saving. Thousands gathered to watch, among them Joseph Mallord William Turner, whose sketches and watercolours of the spectacle have an almost Impressionistic immediacy. With a loud bang and a frenzy of sparks, the roof of the House of Lords caved in, sending flames leaping high into the air. By morning, little was left.

It was almost as if the building had consumed itself in sympathy with the old system. With the benefit of historical distance, the 1830s appear now as a decade

Prolific early Victorian architect Sir Charles Barry was knighted for his work on the new Houses of Parliament.

Augustus Welby Northmore Pugin: working on The Westminster Clock Tower hastened him to his death in an asylum.

of new beginnings in Britain: the Factory Act; the abolition of slavery; the establishment of town councils; the opening of the London-to-Birmingham railway; a new young queen on the throne; and a reformed Parliament, which, eventually, would sit in a modern building, construction of which would occupy much of the first half of Victoria's reign.

The Palace of Westminster was to be rebuilt to the design of Charles Barry. As an assistant, Barry employed Gothic Revivalist and Catholic convert Augustus Welby Northmore Pugin, who, barely out of his twenties, was a rising talent in the neo-medieval movement that gained in popularity in the 1830s.

Significantly, the original plans did not include a clock tower. However, there had been a clock tower at Westminster since the late thirteenth century (all that remains of that original clock is the bell, which now tolls at St Paul's), so there was *going* to be a clock tower in the new Houses of Parliament – and not just any clock tower.

Right from its inception, the new clock was intended to be much more than a public timepiece; it was to be a symbol of Britain's international status, or, as the Office of Works put it, 'a noble clock, indeed a king of clocks, the biggest the world has ever seen, within sight and sound of the throbbing heart of London'.[1] A patriotic tour de force, it was to be one of the marvels of the Victorian age, a time when mankind was unafraid to think big – very, *very* big.

Believing the appointment to be in his gift, Barry asked if court clockmaker
Benjamin Vulliamy could do the job. However, the construction of the clock that
would beat not just at the throbbing heart of London, but a growing global empire,
was taken seriously in other quarters: the Astronomer Royal wrote to Lord Canning,
the chief commissioner of public works, to suggest Edward John Dent for the job.
Canning, in a masterstroke of delegation, asked if Airy would be so kind as to
establish the criteria for the clock, run the tender and deliver a clock that showed
British science and manufacturing at its best.

George Biddell Airy, another 'new face' of the 1830s, had enjoyed a remarkable
career at Cambridge, where he had graduated as senior wrangler, won the Smith's
Prize and, after being elected a fellow of Trinity, was appointed Lucasian Professor
of Mathematics, then Plumian Professor of Astronomy. He had been director of
the Cambridge Observatory until becoming Astronomer Royal in 1835, when,
as mentioned in the previous chapter, he was faced with the Augean task of
reorganising the Royal Observatory along more professional lines. He set to this task
with the industry that characterized the coming Victorian age, introducing many
reforms and setting rigorous standards in every aspect of Observatory life, including
staff punctuality.

He was no less exacting in his criteria for the Westminster clock, requiring the
striking mechanism to be accurate to the second: a standard many thought
impossible. Airy clearly intended to check on the performance of the clock himself:
another condition was the installation of electric equipment to communicate
telegraphically with the Royal Observatory.

*Senior wrangler, fellow of Trinity,
Lucasian Professor of Mathematics,
Plumian Professor of Astronomy and
former director of the Cambridge
Observatory, George Biddell Airy was
appointed Astronomer Royal in 1835.*

The immense, yet minutely accurate, mechanism of the Great Clock of Westminster.

By the time Dent triumphed at Vulliamy's expense, creating bitter resentment, it was a dozen years since the fire that had destroyed the Houses of Parliament. Time was moving on. Among those impatient at the delay was Edmund Beckett Denison, a barrister by profession and gifted horologist by inclination. He wrote to the new chief commissioner of public works to mention as much and, following Lord Canning's example, the new commissioner invited Denison to join the project. Denison consented with alacrity, pored over the plans and pronounced Dent's the best – before suggesting numerous alterations. These, says one of the clock's historians, 'virtually amounted to a redesign', but a redesign that was necessary: 'Had Denison's revisions not been incorporated into the clock's movement, it is unlikely that the prescribed level of accuracy could have been achieved.'[2]

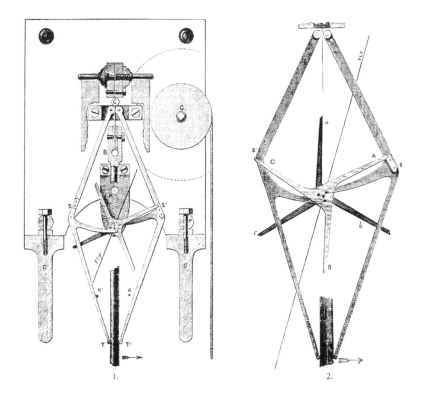

Left: four-legged gravity escapement. Right: double three-legged gravity escapement.

A contract was finally made with Dent in January 1852. February of the same year saw Charles Barry racing against time to complete the design for the clock tower – or rather pressing his gifted but ailing assistant Pugin, now suffering severe mental lapses and delusional spells, to outpace the breakdown that loomed over him. Barry would later attempt to conceal Pugin's contribution to the world-famous tower, but as Pugin's 2007 biographer Rosemary Hill amusingly puts it, the recently knighted 'Barry, who still could not design a door knob in the medieval style, was entirely reliant on Pugin for the conception'. [3]

Shortly after handing over the drawings, Pugin collapsed, and although he rallied and set off to the capital with his son on 25 February, 'by the time Pugin arrived in London he was psychotic'. [4] The Westminster clock tower had, literally, driven him mad. He celebrated his fortieth birthday in an asylum. It was almost as if he had entered into a Mephistophelean pact to design one of the most famous structures in the world at the cost of his brilliant mind. By September, he was dead.

Building the clock was also proving difficult. The bitter Vulliamy had mounted a legal challenge to Dent's appointment. And then there was the problem of the escapement…

Hammer and bell finally in situ high above the Houses of Parliament.

Dent had initially suggested a deadbeat escapement with a constant force device, or *remontoire*, to achieve the required level of accuracy. Among Denison's proposals was a three-legged gravity escapement. It worked well, but Denison felt he could improve it. Next, he created the four-legged gravity escapement, which worked even better. Feeling that there was still room for improvement, he finally came up with the now-legendary double three-legged gravity escapement, which is described by Peter Macdonald, author of *Big Ben: The Bell, the Clock and the Tower*, as 'a truly ingenious piece of engineering'. As he explains:

> *It was designed to be sensitive enough to maintain the required level of accuracy while preventing the effects of external pressures such as driving snow and wind on the hands from being reflected back to the pendulum and thus affecting the clock's timekeeping. So important was its invention that this escapement was considered to be one of the greatest advances in the science of horology: it was soon adopted as the standard and remains unaltered to this day, having been fitted to most large turret clocks throughout the world.[5]*

The entire assembly of bells, wheels, pinions, hands, glass, weights, chains and of course the famous double three-legged gravity escapement is known as the Great Clock of Westminster, but is usually and, incorrectly, called Big Ben. Big Ben is in fact the name given to the largest bell; the eponym of which was either a pugilist or civil servant: Ben Gaunt, a champion prize-fighter, or Sir Benjamin Hall, another new chief commissioner of works who was in post when Big Ben was cast in Stockton-on-Tees in 1856.

Benjamin Hall, first Baron Llanover, eponym of the big bell, by George Zobel (after Frederick Yeates Hurlstone) (1851–1881).

The bell weighed 16 tons and was 9 ft 5 inches across. Too big and heavy for road or rail, it needed to be shipped to London, where it was dragged through the capital on a carriage pulled by eight pairs of horses along a route lined with cheering crowds. Even though more than twenty years had passed since the Palace of Westminster had burned down, the clock tower was still not complete, so the bell was set up at the foot of construction, where it was tested with a hammer weighing half a ton that required the strength of six burly men to wield it. By 1858, the tower was nearing completion, but one morning a huge crack appeared in the bell.

'Poor Ben had to be broken to pieces on the spot where he was first deposited at the foot of that clock tower to which his admirers hoped he was to have given voice for centuries to come,'[6] reported the *Illustrated London News*. Breaking up Big Ben took a week. Then it was back to the drawing board – or, rather, to the foundry. This time it would be the Whitechapel Foundry that was tasked with casting the new Big Ben.

Had they been superstitious, the Victorians might have talked of the 'curse' of Big Ben. In 1852, it had tipped Pugin into insanity then death. A year later, Edward Dent had also died, leaving his stepson Frederick to complete the work. And now the bell itself had cracked.

Another attempt was made. This time a lighter bell was cast and in October 1858, in an operation lasting thirty hours, it was hauled up to the belfry where it joined the quartet of quarter bells. Even so, it was not until 31 May 1859 that the clock first started to work, and it was not with a triumphal peal but in silence, with hands moving around just two of the four faces (the original cast-iron hands were too heavy and needed to be replaced with lighter copper ones). Finally, on 11 July, Big Ben tolled its famous hourly strike, and in September it was joined by the quarter bells. At last, the aural articulation of empire was heard… only to fall silent on 1 October, when Big Ben cracked once again. The curse of Big Ben was apparently alive and well. To top it all, by 1860, both Frederick Dent and Sir Charles Barry were dead.

The solution this time was a stroke of empirical genius: George Airy suggested that the bell was turned through ninety degrees and that the hammer be made lighter. Big Ben sounded again in 1862 and, at the time of writing – forty prime ministers, six monarchs and two world wars later – the same system continues to function. If you look at the hammer closely, you can see that it has become blunt and rounded as, over the course of a century and a half, it has been gradually worn down as the bell takes its revenge.

Casting the bell for the Westminster Clock Tower, 1856. Tapping furnaces at Warner & Sons' Barrett Furnaces, Stockton-on-Tees, England. From The Illustrated London News, *August 23 1856.*

Two decades and more had passed since the Palace of Westminster had burned down but the clock tower was still unfinished, so the bell was set up at its foot, where it was tested with a hammer weighing half a ton.

To visit Big Ben is a powerful experience. The view from the belfry is remarkable; even on a grey London day, the sight of the sprawling city beneath one's feet is not easily forgotten. But the view underneath the bells, in the clock room, is perhaps even more impressive, as the original machinery built by Dent and perfected by lawyer-turned-hobby-horologist Denison continues to function to this day. Made of cast iron, steel and brass, the flat bed construction is almost five metres long and is of such dimensions as to be reminiscent of a traction engine rather than a clock. There are three gear trains: the going train, which drives the hands; the strike chain, which operates Big Ben; and the chiming train, which operates the quarter bells.

Yet, for all its size and presence, it is an extraordinarily delicate mechanism, adjusted to within two-fifths of a second over the course of twenty-four hours by adding or removing a pre-decimal penny to a small tray at the top of the pendulum, thus minutely altering its centre of gravity. When it comes, the striking is rather less impressive than you might think. Indeed, sitting in the clock room, the most remarkable thing is the loud rattling and clattering of the fly fans – giant indoor weathervanes that regulate the striking speed. In 1976, it was the chiming train fly fan that failed, releasing the one-ton weight with such dramatic results that it was thought a terrorist bomb had been detonated. To be fair, the clock room did look like a bomb site, with clock components hurled around the room and through the ceiling. Such was the scale of repairs that the peal of bells was only just ready in time for Her Majesty's Silver Jubilee visit on 4 May 1977.

Over the years there have been minor improvements to the Great Clock of Westminster. For example, behind the milky-white panes of glass that make up each clock face is a bank of electric light bulbs; until the early twentieth century, these would have been gas-lit. On the whole, however, were they to visit today, Airy, Denison, Dent and of course Pugin would find much that was familiar.

Moreover, even though it had been a titanic, decades-long struggle, they could congratulate themselves that they had not only met, but far exceeded the original brief to make 'a noble clock, indeed a king of clocks, the biggest the world has ever seen'. Even kings must eventually vacate their thrones, and the biggest is only the biggest until something bigger comes along – instead, what those long-dead Victorians had created was no less than a national symbol, a historic monument that is to Britain what the Statue of Liberty is to the USA.

What is more, it is an enduring metaphor for the Victorian age itself. Although the exterior is decorated in the ornamental idiom of Pugin's beloved Gothic past; inside, the most ingenious, enduring and at the time most modern technology was used to create a clock that has undoubtedly become the most famous timepiece in the world.

---

1. Quoted in Peter Macdonald, *Big Ben: The Bell, the Clock and the Tower* (Stroud, Glos.: The History Press, 2005), p.18
2. *Ibid.*, p. 23
3. Rosemary Hill, *God's Architect: Pugin and the Building of Romantic Britain* (New Haven, CT: Yale University Press, 2007), p. 482
4. *Ibid.*, p. 484
5. Quoted in Peter Macdonald, *Big Ben: The Bell, the Clock and the Tower* (Stroud, Glos.: The History Press, 2005), p.27
6. *Illustrated London News*, 6 March 1858

The Missed Train Changed Time

Meridian Time

In the summer of 1876, Sandford Fleming was visiting Londonderry (in what was then north Ireland, now Northern Ireland) from Canada. With two days to spare, he decided to see a bit of the countryside. Consulting the 'Official Irish Travelling Guide' along with 'persons resident in Ireland, and accustomed to travel', he devised an itinerary that would enable him to leave Londonderry during the morning and return on the evening of the following day. Fleming was a railway engineer, surveyor and mapmaker, so, as can be expected, his two-day trip was perfectly timed, and he arrived at Bandoran Station at 5.10 p.m. on the second day, in plenty of time to catch the train that the guide 'indicated would leave at 5.35 pm', which would allow him to change to an express train on the mainline and 'reach Londonderry at ten o'clock the same evening.' [1]

Fleming's was a tidy mind, so what he encountered at Bandoran must have shaken him profoundly. He discovered that he had arrived, depending on your view, either 11 hours and 35 minutes late, or 12 hours and 25 minutes early for his train. The guide had been misprinted: 'p.m.' should have read 'a.m.'. 'There was no help for it but to remain at Bandoran until next day,' Fleming wrote, to take the 5.35 *a.m.*, which did not, incidentally, 'like the supposed afternoon train, run to meet an express train on the main line.' [2] Instead of arriving at 10 p.m. on the second day, therefore, he finally pulled into Londonderry Station at 1.30 p.m. on the third day.

Thus inconvenienced, even the most eminent Victorian could have been forgiven for muttering an oath or two. Some may have assuaged their vexation with a glass or two of Irish whiskey. Others may have instructed the station master to bring pen, ink and paper to draft a strongly worded letter to the publisher of the guide. But Sandford Fleming was not a man for half measures; he decided that the world would have to change the way it told the time. There was nothing else for it.

Fleming set out his views in a detailed pamphlet in which he called for the day 'to be divided into twenty-four equal parts, and these, again, into minutes and seconds by a standard timekeeper or chronometer, hypothetically stationed at the centre of the earth'. Had such a hypothetical timepiece existed, and had time been recorded in twenty-four-hour format, Fleming would have been spared his lengthy wait.

Sandford Fleming (1827–1915), who missed his train and changed the way the world told the time. From a collection of images taken for the 15-volume The Pageant of America: A Pictorial History of the United States *(1925–1929).*

'It is proposed that, in relation to the whole globe, the dial plate of the central Chronometer shall be a fixture,' he wrote,

> that each of the twenty-four divisions into which the day is divided shall be assumed to correspond with certain known meridians of longitude, and that the machinery of the instrument shall be arranged and regulated so that the index or hour hand shall point in succession to each of the twenty-four divisions as it became noon at the corresponding meridian. In fact the hour hand shall revolve from east to west, with precisely the same speed as the earth on its axis, and shall therefore point directly and constantly towards the (mean) sun, while the earth moves round from west to east.[3]

Seldom can the wait for a train have yielded such remarkably carefully argued results. The close type of his lengthy essay, a manifesto for the adoption of what he called 'Terrestrial time', covers thirty-seven pages, with diagrams, tables and statistics. Demonstrating an engineer's thoroughness, he seemed to have thought of everything: how to adapt existing timepieces to his twenty-four-hour plan,

Rapid expansion of the rail network led to standardization of the world's timezones. These are two railway workers' timepieces, c. 1900 and 1840. On the left, a Manchester, Sheffield and Lincolnshire Railway pocket watch made by J. Adkins & Sons of Coventry, c. 1900; on the right, a Birmingham & Derby Junction Railway guard's timepiece made by George Littwort of London, c. 1840. NRM photograph taken in 1977.

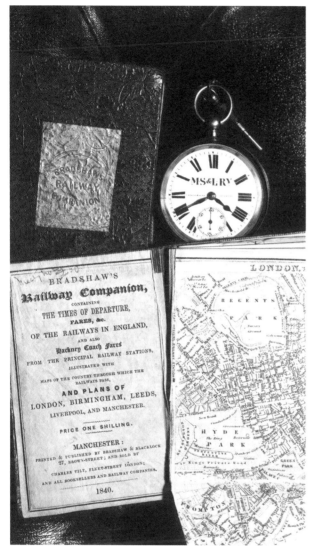

which meridian should be selected to mark the beginning of the universal 'Terrestrial day', and so on. He believed that in the modern world of high-speed rail travel and dazzlingly fast telegraph communication, the twelve-hour method of telling the time was anachronistic, even barbaric.

His conclusion was magisterial:

We have undoubtedly entered upon a remarkable period in the history of the human race. Discoveries and inventions crowd upon each other in an astonishing manner. Lines of telegraph and steam communications are girdling the earth, and all countries are being drawn into one neighbourhood – but when men of all races, in all lands are thus brought face to face, what will they find? They will find a great many nations measuring the day by two sets of subdivisions, as if they had recently emerged from barbarism and had not yet learned to count higher than twelve. They will find the hands of the various clocks in use pointing in all conceivable directions. They will find at the same moment some men reckoning that they live in different hours, others in different days. Is it not important, then, that an attempt should be made to provide a change for this state of affairs and devise for common use some simple uniform system which all nations may with advantage adopt whenever they may feel inclined to do so.[4]

His words were fired with a crusading Esperantist idealism. 'My duty has been simply to attempt to draw attention to the subject, and submit some suggestions for consideration. The subject is one which I feel concerns all countries, although in different degrees, and I shall be gratified if I have in any way assisted in initiating a discussion, which may result ultimately in the production of a matured comprehensive scheme suitable for all lands and advantageous to all mankind.'[4]

'I at first took the national meridian of Washington, and having divided the country into three 15-degree belts, patiently marked out the longitude of some 8,000 stations along some 500 railroad lines': Charles Ferdinand Dowd tackles the Augean task of synchronizing time on America's railways. (American wood engraving, dated 1883.)

Fleming was not the first man to have been struck by the inadequacy of existing methods of time measurement for dealing with the great technological advances of the nineteenth century. As well as ante- and post-meridian confusion, there was the difficulty of keeping track of a highly mobile midday. Solar noon occurred at different times, depending on the location, and by the early nineteenth century there were as many time 'zones' as there were major cities. The Patek Philippe Museum in Geneva has a 'world time Bonbonnière' – a Geneva-made timepiece housed in a circular box engraved with no fewer than fifty-three locations. Somewhat more elegant is an 1804 Paris-made 'Civil Hours' watch, with Paris time in the middle and the time in twelve other cities shown on a dozen individual subsidiary dials, arranged around the edge of the watch-face.

Across the Atlantic, as America underwent explosive economic and industrial development after the Civil War, and cities sprouted from sea to shining sea, the situation became acute. However, in 1870 a solution arrived from a ladies' seminary in the fashionable resort of Saratoga Springs in upstate New York.

Professor Charles Dowd, the principal of the seminary, was another avid temporal pamphleteer who was losing his patience with the way time was kept on the railroads. The task facing him was Augean. Each railway company operated on its own time, usually that of the town in which its head office was located, and with every year time became more tangled. In five decades, the amount of rail track in the USA would grow from twenty-three miles in the early 1830s to over 93,000 in the early 1880s.[5] North America may have been united by a common language, but it was fragmented into a patchwork of dozens of time zones. For obvious reasons, including timetabling and the avoidance of collisions, accurate timing on the railways was extremely valuable – so valuable that railway pocket watches were made with locks that ensured that only the key holder could change the time.

Dowd set about his work diligently.

> *I at first took the national meridian of Washington, and having divided the country into three 15-degree belts, patiently marked out the longitude of some 8,000 stations along some 500 railroad lines, and had a map engraved showing the hour sections and the proposed Standard versus the actual time at each station. With this map was incorporated and published a pamphlet of 100 octavo pages, which I sent to all railroad men and others in this country who would likely feel an interest in the work.*[6]

Nor was it just the growth of rail travel that highlighted the inconvenience of prevailing timekeeping. The progress of science was being held back, since large-scale observed phenomena could not be properly analysed if observers in different locations used varied times to record events.

At the same time as Dowd was petitioning the railroad barons of early Gilded Age America, moves towards the regularization of timekeeping were taking place on the other side of the world. In February 1870, Otto von Struve, director of the Pulkovo Observatory, addressed the Imperial Russian Geographical Society on the selection of a prime meridian. The selection of a globally accepted line of longitude, from which all other longitudes are calculated, was crucial to establishing a standard time.

There were three contenders: the Paris Meridian; the Ferro Meridian (which was essentially the same as the one used by Ptolemy in the second century); and the Greenwich Meridian. Naturally, there was an element of national interest in the question, not least because of the historical rivalry between England and France, but for Struve the issue was purely scientific, as he told the fellows of the Imperial Russian Geographical Society: 'The question of the unification of meridians does not depend on any consideration of political economy, it concerns the scientific world alone.'[7] Accordingly, his choice was Greenwich based on its wide usage on navigational and scientific maps and charts.

The need for a prime meridian was symptomatic of a global change in terms of accurate timekeeping and the closing phase of the age of exploration. Since the days of Columbus, the European powers had been on a protracted landgrab. But by the closing decades of the nineteenth century, as Sandford Fleming wrote, 'all countries are being drawn into one neighbourhood'. More interconnected and interdependent than ever before, the age of the global village was waiting just below the horizon.

Struve was a respected scientist, and if one adheres to a Victorian correlation between extravagant facial hair and eminence, then his splendid long dangling whiskers, of the type known as Piccadilly weepers, commanded attention. His words carried weight, and the issue was aired at the first International Geographic Conference in Antwerp in 1871. (At the time, geography was only just emerging as

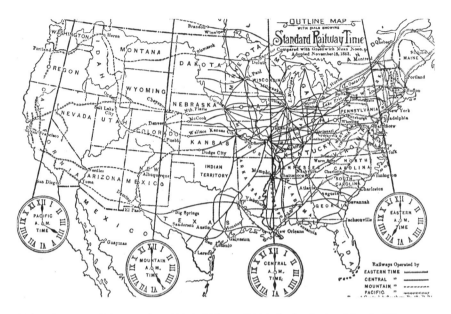

Order at last: a map of time zones into which the USA was divided after the adoption of Standard Time on 18 November 1883.

a codified discrete academic and professional discipline – another sign that society was moving from exploration to understanding.) It was generally agreed that something needed to be done, and yet nothing was.

The subject was discussed once more at the second International Geographic Conference in Paris in 1875, where the host nation did its best to support the Parisian claim to Prime Meridian status. Aside from some intriguing suggestions, such as the adoption of Jerusalem on the basis of its neutrality (absence of an observatory notwithstanding), little more than a continuation of the debate was achieved.

At the next year's conference, in Brussels, delegates' minds were occupied with the division of Africa between European powers, and while the 1878 and 1879 editions featured debate on the subject, again, no decisions were made. The result was that, by the end of the decade, the issue remained unresolved.

Things were not moving nearly fast enough for Sandford Fleming. In fact, they were not moving much at all, so in 1879 he issued another pamphlet, *Time-Reckoning and the Selection of a Prime Meridian to be Common to All Nations*. In 1880, he retired from his work on the railways to devote himself to his time-keeping crusade, and the following year he travelled to Venice to present a paper on the subject at another International Geographical Congress. Crucially, Fleming understood that without a set of clear resolutions and aims, Venice 1881 would turn out to be as impotent as Antwerp 1871 and Paris 1875.

'More than any other delegate at Venice in 1881, Fleming knew that the views of scientists counted for little unless they subsequently made formal representations to their own governments with a view to legislation,' writes meridian historian Charles W. J. Withers. 'There is certainly a sense among the scientists involved that Venice was important. Otto Struve even anticipated its significance.' [8]

Fleming had written to Struve to distribute copies of his pamphlets. He had also started a correspondence with Cleveland Abbe, who in 1879 had published his report 'On Standard Time'. Abbe's interest in standard time was not born out of a missed train in an obscure Irish town, but his role as head of the US Weather Bureau. In order to live up to his sobriquet 'Old Probability', he required a system of timings for readings that was consistent, and so he vowed to 'hammer away at our national congress and call for its action on the subject'. [9] Abbe also wrote to Sandford expressing his hopes that one day the Canadian or American government might host an international convention.

The 'hammering' worked. An act was passed in the USA in 1882 that gave the president power to call a conference to 'fix on and recommend for universal adoption a common prime meridian to be used in the reckoning of longitude and in the regulation of time throughout the world'. This piece of enabling legislation cleared the way for the International Meridian Conference of 1884, which would establish GMT (Greenwich Mean Time) as the global standard.

Between 1882 and 1884, events moved apace. The 1883 International Geodetic Conference in Rome called for governments to select the Greenwich Meridian as the reference from which time should be calculated, and then, just a few weeks later, something miraculous happened: at noon on 18 November, America synchronised its timepieces. Fearing unwelcome government intervention and imposition of an inimical solution that might be prejudicial to their profits, after years of indifference, America's railways had tackled the task of coordinating time. At last, the coast-to-coast cacophony of bells striking noon that rolled across the continent like some aural Mexican wave would be silenced by a system of so-called time belts, as the *New York Times* informed its readers on the eve of the change:

> *One effect of the change of the time standard will be to put an end to the three and one-quarter hours of continuous clock-striking from one end of the Union across to the other. All the clocks in the country, if set correctly, will to-morrow strike simultaneously. But while those in this City and the eastern time belt are striking for 12 o'clock those in New-Brunswick Nova Scotia, and Newfoundland will strike for 1 o'clock; those in Chicago, St. Louis, New-Orleans, and other cities in the central time belt will strike for 11 o'clock; those for Denver and the mountain time belt for 10 o'clock, and those for San Francisco and the Pacific coast time belt for 9 o'clock.* [10]

World Time Ground Zero: the prime meridian line for zero degrees longitude at Greenwich, London.

The paper also described the effects of the change on a rail passenger travelling from Boston to San Francisco:

> *Instead of being obliged to consult conductors, clocks, and time-tables to find the difference between Boston time and the 20 different times he would have encountered yesterday, he will find that the minute band of his watch is always correct and that the difference in time will only be indicated by the hour hand. This will be successively one and two hours late as he moves westward, and three hours late when he arrives in San Francisco. At no time will his watch vary by quarters or half-hours or fractional parts of an hour.* [10]

Struve, had he read the *New York Times* that day, would have been delighted to note that clocks on the east coast of America 'will announce the noon hour, actual time, at the seventy-fifth degree west from Greenwich, England, that point being selected as the unit because it regulates the nautical time of the world'.[11]

The following year the International Meridian Conference in Washington regularized the new temporal status quo across the planet. Rather charmingly, there were still aberrations. France kept its own time until 1898, after which it used the complex formula of 'Paris Mean Time, retarded by nine minutes, 21 seconds', which at least served the purpose of keeping any mention of perfidious Albion out of France's timekeeping arrangements. And it was not until 19 March 1918, with the Standard Time Act, that the established 'time zones' (as they were now known) entered federal law in the USA.

Happily, the outcome of the International Meridian Conference did not bring Fleming's temporal pamphleteering to an end. In 1886, he published *Time-Reckoning for the Twentieth Century*, in which he fulminated and fulgurated against his old foe – the twelve-hour time system.

> *The division of the day into two halves, each containing twelve hours, and each numbered from 1 to 12, is also a fertile source of error and inconvenience.*

> *Travelers who have had occasion to consult railway guides and steam-boat timetables will be familiar with the inconvenience resulting from this cause; none know better by experience how much the divisions ante meridian and post meridian have baffled their inquiries, and how often these arbitrary divisions have led to mistakes. Were it necessary, innumerable instances could be given.*[12]

A decade may have passed since his Irish excursion, but it appears that the passage of time had done little to heal the psychological wounds left by his long wait at Bandoran Station.

As for Dowd, his wounds were physical rather than psychological. He was knocked down and killed by a train in 1904 so did not live to see his plan made law.

1. Sir Sandford Fleming, *Terrestrial Time: A Memoir* (London: 1876)
2. *Ibid.*
3. *Ibid.*
4. *Ibid.*
5. *New York Times*, 20 November 1983
6. Quoted in Derek Howse, *Greenwich Time and the Discovery of the Longitude* (Oxford: Oxford University Press, 1980), p. 123
7. Quoted in Charles W. J. Withers, *Zero Degrees: Geographies of the Prime Meridian* (Cambridge, MA: Harvard University Press, 2017), p. 140
8. Charles W. J. Withers, *Zero Degrees: Geographies of the Prime Meridian* (Cambridge, MA: Harvard University Press, 2017), p. 152
9. Quoted in Ian R. Bartky, 'The Adoption of Standard Time', *Technology and Culture*, Vol. 30, No. 1 (January 1989)
10. *New York Times*, 18 November 1883
11. *Ibid.*
12. Sandford Fleming, *Time-Reckoning for the Twentieth Century* (Montreal: Dawson Bros, 1886)

Time Flies
The Cartier Santos

At about a quarter to three in the afternoon of 19 October 1901, all eyes in central Paris were turned skywards, hoping to catch a glimpse of a cigar-shaped object. Airships had become a familiar sight in the skies above Belle Époque Paris, and suspended perilously below the balloon that autumn afternoon was a short, slightly built, elegantly dressed man, known affectionately as Petit Santos, arguably the city's greatest celebrity 'aeronaut'.

Parisians watched with excitement as, powered by large whirling white canvas propellers, the yellow airship made dignified, but surprisingly rapid, progress from the Aéro-Club de France in St Cloud to the Eiffel Tower, which it circled before commencing the return leg of its journey, in pursuit of the Deutsch prize, which would award 100,000 francs to the first man to complete this aerial circuit in under thirty minutes. As it passed overhead, crowds cheered and men hoisted their hats aloft on the ends of their canes in appreciation.

The commercial value of Santos-Dumont's achievements in the area of powered flight was appreciated early, as this trade card from a chocolate maker demonstrates. (Trade card illustration.)

Alberto Santos-Dumont (1873–1932), Brazilian aviation pioneer. Here in his airship no. 6 descending into a storm of controversy, after his (eventually) successful attempt at the Deutsch prize in 1901. (Inset portrait from Scientific American, *November 1901.)*

Although the return leg of his Deutsch prize attempt was slowed by a headwind and temporary engine failure, the aeronaut arrived over the Aéro-Club twenty-nine minutes and fifteen seconds after he had left. He made a sweep of the airfield, and it was another minute and twenty-five seconds before workmen had caught and tethered the guide rope.

The aeronaut himself did not know the time, however. While he had been hundreds of feet above the streets of Paris, astride what was essentially an automobile tricycle slung beneath a balloon – grappling with his controls, battling the headwinds and at one point even having to restart his engine – he had been in no position to fumble around in his waistcoat, pull a watch out and consult the time.

'Have I won the prize?' he shouted as he neared the ground.

'Yes!' replied the jubilant crowd.

Cult object of the Belle Epoque: the Santos wristwatch.

The Count de Dion, however – one of the judges – looked up from his pocket watch, which he had been studying closely, and said, 'My friend, you have lost the prize by forty seconds.'[1] According to the recently amended rules, the airship should have returned and *landed* within the thirty minutes.

Understandably, the aeronaut thought differently: 'The air-ship, carried by the impetus of its great speed, passed on as a racehorse passes the winning post, as a sailing yacht passes the winning-line, as a road racing automobile continues flying past the judges who have snapped its time. Like the jockey of the racehorse, I then turned and drove myself back to the aerodrome to have my guide rope caught and be drawn down.'[2]

The crowd, too, warmly disputed the judges' decision; in their eyes he had won, and women in particular were swift to show their enthusiasm: 'a number of ladies who were present threw flowers over the aeronaut; others offered him bouquets, and one admirer, to the amusement of the onlookers, even presented him with a little white rabbit.'[3] It became a cause célèbre, with public demonstrations championing Petit Santos and protesting against the judges' decision.

Eventually, after several days of heated debate, he was awarded the prize, but not before the news of his triumph had been reported around the world, with the caveat that there was doubt as to whether he would be awarded the prize money. 'I do not care personally for the 100,000 francs,' he remarked flamboyantly. 'I intended to give it to the poor.'[4]

The money was irrelevant; Alberto Santos-Dumont had more than enough of it, thanks to the fortune generated by his family's sprawling Brazilian coffee plantation on which he was born in 1873 and where he had spent much of his childhood tinkering with the machinery that processed the coffee beans. By the age of seven, he was driving steam traction engines in the fields. Five years later he was in the cab of

A monumental achievement: Santos-Dumont alongside the statue raised in honour by the Aéro-Club of France.

the locomotives that plied the plantation railway. When not operating heavy machinery, he was making aircraft 'driven by springs of twisted rubber', or little 'Montgolfiers'[5] out of silk-paper.

In 1891 he went to Paris, where he bought himself a 3.5 horsepower Peugeot roadster. He was back the following year, when he embraced the new fad for 'automobile tricycles', renting a velodrome for a race between himself and other early motorists. Wheeled vehicles, however, were just a distraction from his real dream: air travel. His true obsession was with flight, and he was disappointed that, even a century after the Montgolfier brothers, there were still no steerable hot-air balloons.

When he returned to Paris in 1897, he made his first ascent in a spherical balloon of a type that had become famous during the siege of Paris. He was enchanted. All the clichés about aviation still had a freshness about them, and he wondered at seeing human beings reduced to the size of ants, and houses that looked like children's toys.

He immediately commissioned a balloon, but to his own design, specifying lighter materials and smaller dimensions. The constructors were convinced it would not get off the ground, but it did (it helped that Santos-Dumont weighed just 110lbs and managed to attain the height of 5 ft 5 only with the help of specially adapted boots). It was the first of a long line of innovative airship designs that pitted him against the established practices of balloon building: the airship that in October 1901 rounded the Eiffel Tower was known as 'No. 6'.

He may have been small, but he was intrepid. In August 1901 he had set off in No. 5, on the same circuit from St Cloud to the Eiffel Tower and back, in an unsuccessful attempt to win the tantalizing Deutsch prize. Having started at 6.30 a.m., nine minutes later he was at the tower, but he was far from happy; he had noticed his

Santos-Dumont aboard his airship No. 9 'La Baladeuse' on the Champs-Elysées. 1903.

balloon was leaking. 'I should have come at once to earth to examine the lesion,' he said afterwards. 'But here I was competing for a prize of great honour and my speed had been good. Therefore I risked going on.'[6]

It was a decision that almost cost him his life.

Santos-Dumont's balloon began to descend and then to drop alarmingly, aiming for the Seine. He thought he had cleared the buildings of Le Trocadero, but 'the half empty balloon, fluttering its empty end as an elephant waves his trunk', slapped against the roof and exploded. The piano-wire rigging caught on the side of the building and Dumont was suspended high above a courtyard, helpless and fearful that at any moment the wires might start snapping. Eventually he was rescued by the fire brigade.

Almost his first thought was that he wanted to get airborne again.

Flight was a drug that seemed to offer the sense of release that others found in pipes of opium: 'one seems to float without weight, without a surrounding world,' he once explained, 'a soul freed from the weight of matter… one is almost loth to see the earth again.'[7] He was addicted and wanted to spend as little time as possible on the ground. Even in his apartment, he dined at a high table from high chairs, requiring the butler to use a step ladder.

Having nearly killed himself, the following day he was looking over plans for No. 6, which in October that year would win him the Deutsch prize.

More than a thrill-seeking, death-defying engineering prodigy, Santos-Dumont was also one of the *beau monde* of the Belle Époque, a sort of honorary member of Proust's gratin. Paris at the start of the twentieth century was the world's pleasure capital: haute couture, cabaret, gastronomy, music, motor cars, art, sex… the very latest and finest of everything was available in the French capital. All in all, it is hard to imagine anywhere more different to Kitty Hawk in North Carolina, where a pair of bicycle-building brothers from Ohio – Orville and Wilbur Wright – were also experimenting with flight, using what looked like very large kites instead of balloons.

As the most cosmopolitan place in the world, Paris was a city of dandies – the tastemaker Boni de Castellane and Robert de Montesquieu (the model for Proust's Charlus and Huysmans' Jean des Esseintes), *inter alia* – and, had he not been an aeronaut, Santos-Dumont would have found minor fame as a well-dressed, well-connected man about town. Dumont saw aviation as entirely compatible with elegant living. His hair parted in the middle and plastered against his skull with pomade, his neck emerging from a high collar, his glossy button boots giving him those valuable extra centimetres of height, he was a perfect dandy in miniature. Indeed, as he became famous for his aerial exploits, there was a fad for his signature chin-scratching shirt collar.

Santos-Dumont moved in the best circles, was friends with the Prince of Wales and was received by the Pope. There is something almost comically Proustian about his 1904 memoir, *My Airships*, in which he drops more names than he does ballast from his balloon. His very first balloon trip ends in the grounds of the Château de Ferrières, country seat of the Rothschilds; on another occasion his airship crashes into the tallest chestnut tree 'in the park of M. Edmond de Rothschild'.[8] Happily, this particular mishap took place near the house of 'the Princess Isabel, Comtesse d'Eu, who, hearing of my plight, and learning that I must be occupied some time in disengaging the air-ship, sent a lunch to me up in my tree, with an invitation to come and tell her the story of my trip.'[9] And, just as winter in Paris threatens to curtail his ballooning, he 'received an intimation that the Prince of Monaco, himself a man of science celebrated for his personal investigations, would be pleased to build a balloon house directly on the beach'.[10]

His must be one of the few books covering technical aspects of aircraft design that includes such lines as 'I promptly extinguished the flame with my Panama hat',[11] when writing of an airborne fire, and begins its account of a loss of equilibrium and sudden descent with the words: 'I was finishing my little glass of liqueur when…'[12]

Indeed, the account of the provisions taken on his first balloon ride sets the tone for his career as an aviator:

> *I had brought up with us a substantial lunch of hard-boiled eggs, cold roast beef and chicken, cheese, ice-cream, fruits and cakes, champagne, coffee, and Chartreuse. Nothing is more delicious than lunching like this above the clouds in a spherical balloon. No dining-room can be so marvellous in its decoration. The sun sets the clouds in ebullition, making them throw up rainbow jets of frozen vapour like great sheaves of fireworks all around the table.*[13]

Flying was an aesthetic and sensory diversion from the futility of human life, and when done the Santos-Dumont way, the inflight service was as good as dinner at Maxim's.

In fact, it was while having dinner at Maxim's that he is said to have had the conversation that would lead to the most profound change in the watch for centuries – a change that continues to define the personal portable timepiece right up to our own times.

He had very nearly failed to win the Deutsch prize because, in part, he had been unaware of the time. While celebrating his victory with a dinner at Maxim's, he mentioned how difficult it was to time his journeys while simultaneously piloting the balloon.

Luckily, the man to whom he said this was Louis Cartier, a third-generation jeweller whose eponymous firm had just moved into glamorous new premises on the Rue de la Paix. Santos-Dumont was a Cartier customer; he had ordered a 'slim gold watch, encircled with rubies'[14] for one of his mistresses, and as Cartier's archives show, between 1904 and 1929 he was a good client. The pages of the old ledgers show purchases including timepieces, jewels for women, jewels for men, such as 'bagues chevalières' (signet rings), hat pins, tie pins, cufflinks and desk accessories. But Louis and Alberto had more than just a customer–retailer relationship. The two men moved in similar circles.

One account tells that they had met at a reception given by Henri Deutsch de la Meurthe, eponym of the prize that Santos eventually won. And, as aviation mania took off (excuse the pun) in the early twentieth century, Cartier's name appeared in the lists of guests at monthly dinners given by the Aéro-Club de France,[15] alongside aviation pioneers including Louis Bleriot, Deutsch de la Meurthe, Léon Levavasseur, the Voisin brothers and, of course, Santos-Dumont.

Both men were in the cultural *avant garde* of a country that, at that time in history, prided itself on huge advances in technology and the arts – from aviation to ballet, radiation to painting – and bequeathed to posterity the cinema and the inflatable tyre. During the first decade of the century, Louis Cartier was experimenting with abstract shapes and designs that would later become known as Art Deco. Santos-Dumont was also a man ahead of his time. Indeed, there is a suggestion of Futurist painting about

Ever elegant: Santos-Dumont maintained impeccable standards of dress whether on the on the ground or in the clouds. (Santos-Dumont in his boat, from Revista Moderna *No. 30, April 1899)*

his lyrical description of flying over a city at night: 'We see a point of light far on ahead. Slowly it expands. Then where there was one blaze there are countless bright spots. They run in lines, with here and there a brighter cluster. We know that it is a city.'[16]

The solution to Santos-Dumont's timing dilemma was appropriately daring and futuristic: Cartier moved the watch from the fob-pocket to the end of the arm.

He made his friend a small, straight-sided, curved-cornered watch about the size of a postage stamp that could be fastened to the wrist by means of a leather strap and consulted with just a flick of the wrist.

There had been wrist-worn watches before the twentieth century, and there are various instances of women wearing timepieces attached to bracelets. Queen Elizabeth I is said to have owned one. In June 1810, Napoleon's sister Caroline Murat, Queen of Naples, placed an order with Breguet, Marie Antoinette's favourite watchmaker, specifying a repeater watch for a bracelet. And in 1868 Patek Philippe made its first wristwatch for the Hungarian Countess Kocewicz, who wanted to wear a tiny watch set in a gold bangle.

In extremis, men had also worn time on the wrist: during the Battle of Omdurman in 1898, British soldiers had worn pocket watches in cup-like wrist bands. However, with what he made for his friend Santos-Dumont, Louis Cartier created the first timepiece designed expressly for wear at the end of a man's arm, rather than in his waistcoat pocket. It was an invention that could have only caught on in Paris at that time, and benefited from a concatenation of unique circumstances: Cartier's flair for innovative design; the brand-new need presented by the birth of aviation; and Dumont's reputation as a leader of fashion.

*An unsuccessful attempt at the Deutsch Prize left
Santos-Dumont's life quite literally hanging by
threads. (From* Le Petit Journal.*)*

Wristwatches were considered
effeminate and would continue to be
regarded as such until after the Great
War. But dapper Santos-Dumont, whose
courage was beyond question, did not
feel his masculinity imperilled by a
watch on the wrist. In fact, he already
wore jewellery on his wrist. A female
admirer had presented him with a
medal of St Benedict that she suggested
he wear on his watchchain, in his card
case or around his neck. He chose to
wear it more prominently: 'newspapers
have often spoken of my "bracelet",
he wrote, but 'the thin gold chain of
which it consists is simply the means
I have taken to wear this medal, which
I prize.' [17]

Apparently, it took some time for
Cartier to produce the watch; 1904 is
widely accepted as the date of creation. Although, as Santos-Dumont's biographer
Nancy Winters makes clear, 'because it was a gift, it was never entered on the Cartier
register, so the exact date of making of the Santos is uncertain'. She concedes,
however, that it would have been 'sometime after October 1901 and before
November 1906'. [18] In 1908, Cartier's archives show that Santos No. 2 was made and
sold. Cartier would go on to make many different sorts of wristwatch, but the Santos
was the only man's watch to be named after a client.

Inevitably, flaneurs and men of fashion around Paris asked Cartier to make them
wristwatches too, and, according to Cartier, aviators including Roland Garros
and Edmond Audemars wore the 'Santos' watch in honour of its eponym. Cartier
had 'invented' the man's wristwatch, one of the defining objects of the twentieth
century, and within two or three decades, the pocket watch would become a
museum piece.

If delivered in autumn 1906, the watch would have found its eponym in particularly
good spirits, as by then he had moved from motorised balloons to fixed-wing aircraft,

1904

Cartier, the design genius and aviation enthusiast who created the revolutionary wristwatch for his friend Santos-Dumont, and later sold it under his name. (Portrait by Émile Friant, 1904.)

and that year scooped the Aéro-Club de France prize for the first powered flight of over 100 metres. He had become the most famous man in the world. His aeroplane had made the first publicly confirmed, manned, powered flight and, given the secrecy with which the Wright brothers conducted their early flights, it is still contended by some, unsurprisingly many Brazilians, that Santos-Dumont made the world's first aeroplane flight.

Over the years, Santos-Dumont's name faded from the public imagination. A man who enjoyed truly global fame as the pioneer of aviation, he was gradually upstaged by Bleriot, then Lindbergh and, eventually, Neil Armstrong, as man flew ever faster, higher and further.

'Whether he technically made the first ever heavier-than-air flight is, in a way, irrelevant,' wrote Nancy Winters, 'because for several years the entire world believed he had, and was inspired by his spirit which, in the end – and at the end of the millennium – may be the thing that will never be matched.'[19]

And even if he did not make the first heavier-than-air flight, there is no doubt that Santos-Dumont was responsible for inspiring the world's first, purpose-built man's wristwatch.

1. *Saint Paul Globe*, 20 October 1901
2. Alberto Santos-Dumont, *My Airships: The Story of My Life* (London: Grant Richards, 1904), p. 198
3. *Saint Paul Globe, op. cit.*
4. *Ibid.*
5. Alberto Santos-Dumont, *My Airships, op. cit.*, 21
6. *Ibid.* p. 171
7. *Ibid.* p. 58
8. *Ibid.* p. 167
9. *Ibid.*
10. *Ibid.* p. 213
11. *Ibid.* p. 110
12. *Ibid.*, p. 35–6
13. *Ibid.*
14. Gilberte Gautier, *The Cartier Legend* (London: Arlingon Books, 1983), p. 95
15. *L'Aérophile* (various issues)
16. Alberto Santos-Dumont, *My Airships, op. cit.*, 58
17. *Ibid.* p. 168
18. Nancy Winters, *Man Flies: The Story of Alberto Santos-Dumont, Master of the Balloon* (London: Bloomsbury, 1997), p. 148
19. *Ibid.*

The First Sports Watch
The Jaeger-LeCoultre Reverso

There can be few tales from the Raj that involve a Swiss denture maker, but in 1930 a businessman by the name of César de Trey was in India watching a polo match when he overheard players complaining that the glasses covering the dials of their wristwatches were constantly becoming damaged while playing.

Trey had made his name and fortune in the manufacture of gold and porcelain dentures. According to his 1935 obituary in *Dental News*, he was 'one of the most conspicuous and picturesque figures in the dental trade of Europe during the past thirty years'[1].

As well as embellishing the world of dental prosthesis between the First and Second World Wars, Trey was a horologist manqué, and by the late 1920s he was able to open a small watch business in Lausanne. He entered the trade at a time of immense change: namely, the personal timepiece was in the middle of its traumatic move from waistcoat pocket to the end of the arm. Initially, 'wristlet' watches had been considered an affectation, but their widespread use in the trenches had earned them a following, and by the early 1930s they were bringing the hegemony of the pocket watch to an end.

However, supporters of the pocket watch believed that a timepiece was simply too delicate to leave on the wrist – a view supported by Trey's observations on the polo field, where broken watchglasses were seen as an occupational inconvenience of the vigorous game of polo. Crystal was simply too fragile to withstand balls, sticks, hooves and the other hazards of the polo field; the metal used for the rest of the case was far more robust and would have been the perfect choice, but for the lack of transparency. Trey realised what was needed: a watch that could protect itself – a watch that, in the words of patent application No. 712868, filed at the French Ministry of Trade and Industry on 4 March 1931, 'can be slid in its support and completely turned over'[2].

The subsequent watch was the result of collaboration between Parisian instrument-maker Edmond Jaeger and Swiss movement-maker Antoine LeCoultre ('Jaeger-LeCoultre'). Jaeger engaged designer René-Alfred Chauvot to render the idea practical, and he duly designed a rectilinear watch in perfect harmony with the streamlined, hard-edged Art Deco aesthetic. It was a watch of two parts: a bottom plate or chassis, with tapered lugs to which the strap was attached; and the watch head, which was attached to the chassis using a system of recessed pivots running in

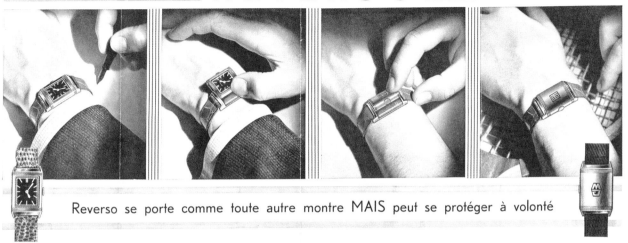

POUR LES HEURES TRANQUILLES **REVERSO** POUR LES HEURES D'ACTIVITÉ PHYSIQUE

Reverso se porte comme toute autre montre MAIS peut se protéger à volonté

The Reverso: conceived on the polo fields of the Raj between the wars and worn around the world today. (1931 Jaeger-LeCoultre Reverso Wristwatch advert.)

guide slots, enabling the watch to be turned over, with a pair of spring-mounted ball bearings with corresponding recesses to snap the watch firmly into place.

It was a system as pretty as it was practical; in addition to protecting the glass of the watch, when turned over it presented a surface fit for engraving and enamelling. At that time, rather fittingly, given its origins, some of the most magnificent examples of decorated 'Reversos' were commissioned by the Maharajahs of India.

Rather more sober was the steel example from 1935, somewhat crudely enamelled on the back with the letters D MAC A, that surfaced in 2015. Owned by General Douglas MacArthur, known for his corncob pipe, aviator sunglasses and ultimate victory over Japan in the Pacific Theatre of the Second World War, its military provenance pushed the price to 87,000 Swiss francs, making it the most valuable Reverso yet sold at auction.

But, in general, the Reverso remained a Jazz Age phenomenon. MacArthur's patronage notwithstanding, its associations with polo players, maharajahs, lounge lizards, bons vivants and boulevardiers did not suit it for military service, and by the end of the war it was almost forgotten. After a few years of popularity, the watch fell from fashion and into a coma: hibernating in the archives for half a century until it was discovered and put back into production, rescuing the firm of Jaeger-LeCoultre from

Pipe-smoking General Douglas MacArthur took advantage of the opportunities for personalization offered by the reversible case design.

what had appeared to be certain extinction during the era of battery-powered and quartz-regulated watches.

Its importance to one of the great names in horological history aside, what guarantees the Jaeger-LeCoultre Reverso its niche in history is its impact on man's ability to keep accurate time no matter what activity he was undertaking. It can be argued that the Reverso was the first sports watch – its functionality created to overcome the

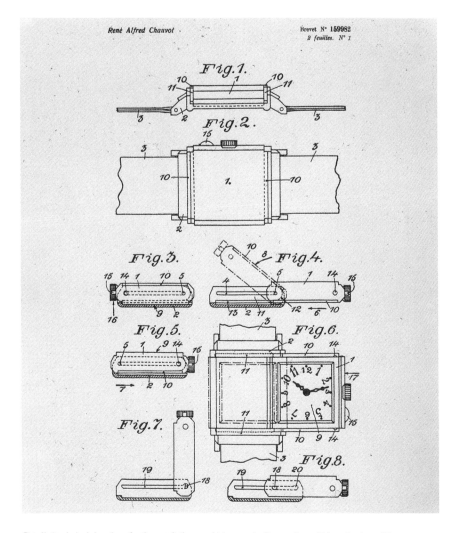

Detailed technical drawings for the watch that would become the Reverso but which at the time of its patent application was rather less catchily known as a watch that 'can be slid in its support and completely turned over'.

demands made of it in the highly competitive environment of the polo field, where stick, ball, hoof, or indeed a fall to the ground, could easily destroy a normal watch.

Of course, had he been of a more banal turn of mind, less 'conspicuous and picturesque', Trey might simply have suggested that the riders take their watches off while playing polo. But had he been of such a pedestrian turn of mind, the world would have been denied a defining and influential design that did much to establish the sports wristwatch.

1. *Dental News*, Volume 18, p48, 1935

2. Manfred Fritz, *Reverso – The Living Legend* (Jaeger-LeCoultre, Edition Braus, 1992), p. 28

The Most Expensive Watch in the World

The Patek Philippe Henry Graves Supercomplication

Mid-November in Geneva is the time of the watch auctions. For one week, the city fills with collectors and dealers from all around the world. Geneva's bars and restaurants buzz and hum with talk of timepieces. Experienced eyes squint through jeweller's loupes, registering every hallmark and serial number; they pore over dials, hands and tiny-toothed wheels for forensic signals that will inform their bids when the time comes to take their seats in the saleroom.

It is a busy time, with potential buyers rushing from viewing to viewing, but in 2014 one auction in particular commanded attention above all others.

On 11 November, Sotheby's would sell 368 lots, including pocket watches and wristwatches made by the great names of horology: Patek Philippe, Rolex, Cartier, Vacheron Constantin and Breguet, *inter alia*. Spanning a period of approximately 200 years from the late eighteenth to the early twenty-first century, and including timepieces as diverse as triangular Masonic pocket watches and solar-powered table clocks, some lots were offered with no reserve, while others came with four-, five-, and six-figure estimates. In short, the catalogue promised the usual polyglot offer of the major auction house sale: an encyclopaedic horological buffet with something to suit almost every budget and taste.

After busy morning and afternoon sessions in one of the reception rooms of the Beau Rivage hotel, the auctioneer returned to the rostrum for the early evening session at 6 p.m. He did so in front of a packed room that almost crackled with a palpably febrile sense of expectation. As well as the familiar faces of collectors, dotted here and there were senior industry figures, owners and chief executives of the major horological marques.

The auctioneer moved briskly through the lots: some Cartier dress watches, a handful of steel sports Rolexes and, with lot 344, a perpetual calendar moonphase minute repeater by Patek Philippe, sold for a respectable 329,000 Swiss francs. He paused, took a sip from his glass of mineral water, moistened his lips and looked around the room. Speaking slowly and clearly, he announced the next lot:

The sidereal (rear) dial shows the star chart over Manhattan, hours and minutes of sidereal time, the equation of time and the times of sunrise and sunset.

'Lot number 345. The extraordinary and very, very important Henry Graves Junior Supercomplication.' After pausing for theatrical effect, he continued: 'I can open the bidding at nine million Swiss francs.'

Time is experienced subjectively, and the man standing at the Sotheby's rostrum in a large private room in Geneva's Beau Rivage hotel on 11 November 2014 was about to experience the longest twelve minutes of his life.

The mean time (front) dial shows the moonphase and calendar functions, power reserve indicators for both the striking mechanism and the movement, the split second chronograph and the hand for the alarm.

This was the event for which the room had filled: nothing less than the sale of the single most important portable personal timepiece to come onto the market for a decade and a half. Not since 2 December 1999 had the world of vintage watches witnessed a similarly important sale. On that day, the very same watch had fetched an astonishing $11 million at auction. Now it was back on the market. The projected price was terra incognita. It was already the most expensive watch in the world; the question concentrating minds that November evening was whether it would eclipse that earlier record. Adding to the drama, 2014 was the 175th anniversary of the firm that had made it, and Patek Philippe president Thierry Stern was sitting in the middle of the packed auction room.

Known simply as 'the Graves', the Patek Philippe Supercomplication, listed a trifle prosaically in Patek Philippe's time-stained ledgers as 'No. 198 385', is a behemoth of a watch. Weighing over half a kilogram, it comprises 900 components, the interaction of which results in two dozen horological complications, including a moving celestial map depicting the changing night sky over New York.

Its first owner and eponym, the son of a successful Wall Street financier, was an otherwise unremarkable member of early twentieth-century USA's leisure class. Henry Graves Jr came accessorized with all the appurtenances of the social elite of Edith Wharton's New York, right down to the family crest depicting an eagle rising grandiosely out of a coronet and a fondness for being photographed on horseback or in riding clothes. Something of a stickler for correct form, he insisted on wearing a suit and tie... even when canoeing.

Under mean time dial, view with calendar
mechanism removed.

Under sidereal time dial, view with
sidereal time train removed showing the time
and chronograph mechanisms.

Under mean time dial, view of calendar mechanism
and moon phases.

Under sidereal time dial, view showing mechanism
for star chart and sidereal time.

Like others of his class he considered himself a connoisseur, collecting paintings, mezzotints, nineteenth-century paperweights and watches… *especially* watches. His life may have lacked the drama and incident of that of, say, Marie Antionette, but, horologically speaking, the watch that has immortalised him is comparable to the one completed by the firm of Breguet, intended for that tragic Queen of France a century earlier. And, significantly, unlike the Marie Antoinette – by November 2014, returned to the Jerusalem museum from which it had been stolen – this watch was on the market.

The Graves has presence and gravitas. Weighing 535 grams, measuring 74 mm in diameter and 35 mm in height, it fills the hand impressively. Its 23.25 cm circumference bristles with slides and push pieces that activate its many and varied special functions. Instead of a front and back, it boasts two faces: one displaying the hours, minutes and seconds of mean time; the other showing the hours, minutes and seconds of sidereal time. In addition, the Graves displays the times of sunset and sunrise and the equation of time.

Its perpetual calendar accommodates years of varying length and shows the day of the month, the days of the week, the months of the year, phases and age of the moon and, of course, the appearance of the celestial vault over its owner's home town of New York. Its split-second chronograph (a stopwatch-like function) alone is a masterpiece, with thirty-minute and twelve-hour totalizers (to record the time passed).

However, it is the audible functions that really distinguish this extraordinary timepiece: a minute repeater with Westminster chimes, *grande* and *petite sonnerie*, plus an alarm function. To hear its crystalline chiming of the hours, quarters and minutes is like listening, in miniature to the bells of a country church pealing out over a silent winter landscape… except that country churches tend not to be equipped to play the melodies sounded by the bells of the Great Clock of Westminster – popularly, if erroneously, known as Big Ben. A quartet of rather more technical functions relating to the going and striking trains and the winding and setting bring the total to twenty-four complications in addition to the regular timekeeping functions.

But the Graves is much more than a highly impressive gadget, a must-have executive toy for the early twentieth-century plutocrat. Although decidedly large for a watch, it is relatively compact when viewed as the crystallization of a discrete cultural period. Mark Twain sarcastically called it the Gilded Age, and the name stuck to the period of industrial and economic expansion between the end of the American Civil War and the beginning of the 1930s that made the USA the world's foremost power. The national wealth of the USA quadrupled between 1870 and 1900, and with around 80 per cent of Americans making under $500 a year, Croesan wealth was concentrated in the hands of a few tycoons who became richer than the royal and imperial families of Europe. The fortunes of the time were truly staggering, and for those who accumulated them, a complicated Swiss watch was one of the axioms of success. The richer the man, so the rule generally went, the greater his interest in timepieces.

Presentation watches marked life's important moments. This watch, made in 1891, was presented to J. Frederic Tams by J.P. Morgan as a souvenir for building the yacht 'Corsair'.

Business for the best Swiss watchmakers boomed. The 1880s saw massive growth in the manufacture of Swiss chronographs, as racing took off as an aspirational pastime. No student of the turf considered himself properly equipped without a precise Swiss chronograph to pull out of his waistcoat pocket and time his thoroughbred's performance on the course at Saratoga Springs in upstate New York or Churchill Downs in Kentucky.

One such example of a timepiece as a jewel to adorn the crown of commercial success was a handsome Patek Philippe minute repeater with split seconds, chronograph No. 90455. Made in the early 1890s, it was owned by distiller Jasper Newton – the proud liquor magnate – who went as far as having the cover of his watch engraved not with his name but that of Jack Daniels (after the product on which his fortune was founded).

Typically, the young princelings of the USA's mercantile elite had their arrival into manhood consecrated with the almost ritual presentation of a complicated Swiss watch. In 1893, Cornelius Vanderbilt Jr was given a richly engraved Patek Philippe minute repeater chronograph with split seconds as a twenty-first birthday present by his father – the famously foul-mouthed 'Commodore'.

Such timepieces were occasionally distributed as marks of special favour to subordinates who had rendered particularly valuable service. In 1901, the steel tycoon Henry Clay Frick took delivery of a handsome chronometer with detent escapement and a one-minute tourbillon, engraved with the image of the Frick Building in Pittsburgh; he presented it to the architect Andrew Peebles.

Unsurprisingly, J. P. Morgan was one of the highest-profile Gilded Age horophiles. Morgan spent on a Wagnerian scale; he collected almost anything, and given that there was so much stuff to purchase and relatively little time in which to purchase it, he would buy up entire collections, acquiring his status-conferring possessions in bulk. It was in this way that he bought the famous Marfels Collection of Renaissance and enamelled watches.

When it came to contemporary watches for everyday use, Morgan favoured complicated pieces from the English maker Frodsham. When he commissioned three identical Rolls-Royces built during 1911 and 1912, each featured Frodsham clocks (and electric cigar lighters). New partners at his bank could expect to be presented with a handsome Frodsham open-face tourbillon with minute repeater and split-second chronograph. Between 1897 and 1926 (the custom was perpetuated by Morgan's son after his death in 1913), twenty-five such 'Morgan-calibre' watches were presented, suitably engraved, to close friends and colleagues. And if they were invited sailing by Morgan, their watches enjoyed special treatment: the staterooms of his yacht, the second to bear the name *Corsair*, were fitted with specially designed niches in which guests could place their watches at night.

Of course, the automotive pioneers of the time, the tech giants of their day, had a natural interest in complex mechanical objects, whether cars or watches. The Dodge brothers were clients of the venerable house of Vacheron Constantin, while during the first quarter of the twentieth century James Ward Packard became one of the most loyal and exigent clients of Patek Philippe. During the first quarter of the twentieth century, Packard ordered thirteen important watches from Patek Philippe, including such esoterica as a watch with a ship's striking mechanism, which sounded every half-hour ending each four-hour 'watch' by striking eight times – the first time this had been accomplished by a watch. Another played a piece from *Jocelyn*, one of his mother's favourite operas. However, the *pièce de résistance* was No. 198 023, which he received in April 1927. It was equipped with ten complications, including a celestial map of the night sky above Packard's hometown of Warren, with 500 stars of varying size picked out in gold against a deep royal-blue background.

Given the similarity of their two watches, for a long time it was considered that Packard and Graves were rivals in collecting, engaged in a contest to own the world's most complicated watch; it is an attractive idea, but alas probably untrue. The two men came from different strata of society, different parts of the country and most likely never even met. Wonderful though Packard's watch is, it is safe to assume that he would have gladly exchanged it for the sight of the real night sky; when he received it, he was at Cleveland Clinic Hospital being treated for the cancer that would kill him within a year, whereas Graves was to wait many years for his watch.

Nevertheless, without the Packard, there would never have been the Graves.

Its realization was a colossal undertaking for Patek Philippe, requiring five years of work before it was completed on 28 September 1932 and delivered to the man who became its eponym the following year at a cost of 60,000 Swiss francs (approximately $15,000–16,000 today). Manufactured to minute tolerances and requiring precise mathematical calculations (some gearing rations are calculated to ten decimal places), it remains the most complex and complicated watch created before the introduction of computer-aided design technology. At that time nothing so ambitious had been attempted within the claustrophobic confines of a watch. In order to create the piece, the firm had to draw upon specialists beyond Patek Philippe's historic headquarters on the Rue du Rhone, not just within Geneva, but from right across Switzerland.

Its construction and subsequent fame would do much to establish Patek Philippe as the world's foremost watchmaker, but it was also destined to be the last in a breed of watches that became extinct on Tuesday 29 October 1929, when the American stock market collapsed. Within a week the market had lost $30 billion from its value. From

Henry Graves: but for the Patek Philippe watch that bears his name, banker Henry Graves would be untroubled by posterity.

stratospheric financial heights, the world dived into the Stygian depths of the Great Depression – a depression that would only end in the six horrific years of the Second World War. It would be decades before the spirit of confidence and prosperity would return, and when it did the world was a changed place. The pocket watch was no longer a status symbol; in the atomic age, it was a museum piece, a historic item that spoke of a long-vanished past. Once an item of technological wonder, it had been reduced to a mere heirloom.

After eighty-six years of blameless affluence, Graves died in 1953 and bequeathed the watch to his daughter, who in turn gave it to her son. Six days after his mother's death, he sold it in March 1969 to an eccentric Illinois industrialist called Seth Atwood

COMMERCIAL MIGHT *VERSUS* **DIVINE RIGHT.**
The Modern Trust King Brings Dismay to the Old Kings of Europe.

'Commercial Might versus Divine Right'. J.P. Morgan symbolized the late nineteenth and early twentieth century American tycoons who became richer than the royal and imperial families of Europe, and for whom a complicated Swiss watch was a sine qua non *of success.*

for $200,000. Atwood had the idea of creating the world's greatest collection of timepieces and, conforming to the stereotype of the rich American abroad, he came to the salerooms of Europe to buy the best watches, heedless of cost.

As far as he was concerned, the Graves was the best of the best. The following year Atwood opened his Time Museum in Rockford Illinois, with the Graves as the star attraction, and there it remained until the museum closed in March 1999 and the sale was entrusted to Daryn Schnipper of Sotheby's. As soon as she heard the news of the museum's closure, she headed to the Midwest to value the collection and negotiate its sale.

Schnipper's association with the watch had begun fifteen years earlier, when Atwood had asked her to value it. Today the elegantly bobbed Schnipper is one of the legends in the world of watch auctions, and that legend began in 1986, when, after five years in the watch and clock department of Sotheby's in New York, she was sent to Rockford to value the Graves. Over thirty years later she well remembers the thrill of first handling the watch which by that time had already accumulated its share of fame as a hallowed object.

She describes an experience of religious intensity: 'It was awe-inspiring. It just radiates magnificence in every sense of the word. I was speechless. I didn't know what to say. It was a life-changing experience in the sense that just handling it communicated how important and how wonderfully built it was. I can't think of enough adjectives. It just awakened all your senses.'[1]

The Graves has the power to make memories; it imprints itself on the minds of those who see it and handle it – let alone sell it. 'It was just mind-boggling to think that the Graves could actually be offered at auction,'[2] recalls Schnipper. Other auction houses were also invited to tender for the sale: 'So we said three to five million. No one had ever put three to five million dollars on a watch and when we came to the day of the sale we weren't sure what was going to happen.'[3]

'We had six bidders up until five million, and that is when I started breathing again!' The tension was immense. 'And then it just kept going and it was two people, with Patek the under-bidders.'[4]

A decade and half later, the watch was back on the market, under curious circumstances. Its owner, Sheikh Saud bin Mohammed Al-Thani, cousin of the Emir of Qatar, had been on a protracted collecting spree that had earned him the sobriquet 'the world's biggest art collector'. Beyond art, he also collected vintage cars, penny-farthings and Chinese antiques. 'Sheikh Al-Thani wasn't a big deal in art buying circles. He was massive,' observed BBC arts editor Will Gompertz in November 2014. 'When he was in town – so the rumour goes – art dealers and

The Graves has the power to make memories, it imprints itself on the minds of those who see it and handle it.

auction houses would dust down their best stuff, add a nought or two, and await his visit.'[5]

By the end of the twentieth century, the world's most extravagant collectors were no longer men such as Seth Atwood. The early twenty-first century saw the art and collectibles market pushed to new heights by money from parts of the world that had not been remotely active on the scene during the late 1960s: Russia, China and, of course, the Middle East. Since the oil shock of the early 1970s, immense financial resources had been concentrated in the hands of a small number of ruling families in the Middle East and, like their American predecessors a century before, they had to have the best of everything… including watches. These financial resources were of a scale unseen before, but they were not entirely limitless, and by

2014 Al-Thani was in financial trouble. He needed to liquidize some of his assets, including the Graves.

During the fifteen years following its acquisition by Al-Thani, the Graves had become a rockstar, with Schnipper as impresario. 'We toured it. We went to Hong Kong, China, New York and all sorts of places that I can't even remember. By that time it was iconic. It was very symbolic of the watch world.'

The power of the watch had magnified, and by the time it came for sale in Geneva, the Graves was no longer a curiosity; it had shouldered its way to the front of the cultural consciousness and, although the catalogue discreetly omitted an estimate, it was rumoured that Sotheby's was seeking a minimum of $15.6 million; which, if achieved, would put a Patek Philippe pocket watch on the same level as paintings by the masters of modern art.

'And then of course it was further complicated by the owner's sudden demise the night before the sale.'[6] At the ripe age of forty-eight, Al-Thani had departed his life of fine art, penny-farthings and supercomplications.

'That was great,' says Schnipper in a voice heavy with irony, 'but I can tell you Sotheby's writes wonderful contracts.'

And so, at around 6.30 p.m., a twelve-minute financial duel began. The skill of the auctioneer lay in prolonging that test of financial strength and irrational desire, bringing it to the brink of conclusion to coax another bid that rose in increments of 500,000 Swiss francs. More than once he raised the gavel above his head and intoned with sacerdotal gravity, 'Fair warning… the gavel is up… last chance… final warning', only to be rewarded with a collective gasp and ripple of applause from the room as, with a last-second bid, the price climbed higher and the watch remained in play. As the price moved past 20 million Swiss francs, two tenacious collectors continued to fight it out in an atmosphere akin to that of the deciding game of a Wimbledon final on centre court. When at last the gavel really did come down and the room exploded with applause, the winning bidder had paid 23,237,000 Swiss francs.

At the time of writing, it remains the most expensive watch ever sold.

---

1. Interview with the author, January 2019
2. *Ibid.*
3. *Ibid.*
4. *Ibid.*
5. 'Qatari art collector Sheikh Saud bin Mohammed Al-Thani dies', BBC News, 11 November 2014 (https://www.bbc.co.uk/news/entertainment-arts-30001716)
6. Interview with the author, January 2019

Jet Set Go
Rolex GMT-Master

*H*ad you been in Seattle on 15 July 1954 and looked skywards shortly after 3 o'clock in the afternoon, you might have glimpsed something rather unusual streaking through the sky: the canary, tan- and cream-coloured Boeing 367-80, a new kind of aircraft with swept-back wings from which hung pods holding engines that did not need propellers.

This was the first test flight of the prototype for what would become the Boeing 707, an aircraft that would change the world. Four years later, it was put into service by Pan American World Airways, more commonly known as Pan Am.

Boeing 367-80 Prototype in flight, when it entered production this aircraft was given the now famous designation '707'.

After a false start (when the British de Havilland Comet had commenced commercial flights in the early 1950s, only to be grounded in 1954 after a series of unexplained crashes), the jet age had begun.

Seventy years earlier, in 1884, the International Meridian Conference in Washington had tidied up the way the world told the time. Now mankind faced a different problem. With the arrival of the jet aircraft, the world's time zones were crossed so rapidly and frequently that it was easy to forget which one you were in. Those piloting these new jet airliners wanted a watch that could inform them of time in two zones simultaneously, at a glance.

It was to meet this new and hitherto unimagined need that in 1955 Rolex presented a new type of wristwatch that looked just as eye-catching as the new jet airliners. In addition to the hands that told the hour and minutes, there was a hand that made one revolution every twenty-four hours and, just as the Boeing 367-80 had attracted attention with its brown and yellow paintwork, so the eye was drawn to the Rolex's blue and red bezel, bearing the numbers 1–24, that rotated around the watch face. It had, so the company's brochures carolled, been 'created by Rolex to meet the very special and exacting needs of the flying personnel of two world renowned aviation companies'. [1]

It was a strange-looking watch, and initially it was marketed in the way that one might have sold a compass or a slide rule. A rather joyless and wholly un-aspirational photograph of a radio operator graced the front of a brochure to introduce the new watch, while advertisements used the slogan 'Navigators navigate with Rolex – The Rolex GMT-Master' [2] over a photograph of a cargo ship.

But with a Rolex GMT on the wrist, navigation had never been so chic: it was the official timepiece of Pan Am pilots and the unofficial watch of a new social elite known as 'the jet set'. Jets shrank the world. An ocean liner would take the better part of a week to steam across the Atlantic; on a jet airliner the journey could be made in little more than an afternoon. The airline pilot was no longer the operator of a lumbering, noisy piece of machinery, but a herald and hierophant of the modern age, clad in Ray-Ban Aviator sunglasses and gold braid. The Rolex GMT-Master was a new watch for the new era.

Air travel had a magic to it, liberated from the ground and suspended between heaven and Earth; hurtling through the skies at over 500 mph, it was hard for passengers not to feel ever so slightly touched by the hand of God. Airports were the churches of a new religion; that of the glamour of high-speed jet travel – a religion of which Frank Sinatra was an enthusiastic proselytizer.

In 1958, at the height of his fame and popularity, Sinatra released his fourth long-playing record. Called *Come Fly With Me*, the sleeve showed Sinatra standing on an

The genius of the Rolex GMT Master lay in its simplicity: a rotating twenty-four-hour bezel and a twenty-four-hour hand were all that was needed to transform a wristwatch into a powerful tool for, among others, airline pilots.

Wearing a Rolex GMT Master was the next best thing to being a pilot: 'Pan Am flies with Rolex' was the proud boast of early advertisements for the GMT Master.

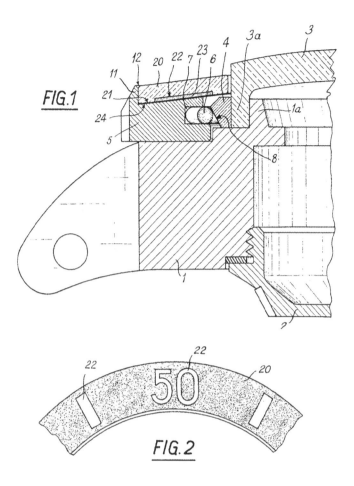

FIG.1

FIG.2

A cross-section of the rotating bezel that made the Rolex GMT master such an effective in-flight instrument, from the Swiss patent office application document dated 28 January 1957.

airport runway with a TWA[3] airliner behind him. It was not so much a pop record as a travel brochure set to music. The title track, 'Come Fly With Me', with its evocations of Acapulco and Peru, was a hymn glorifying jet travel and suggesting its aphrodisiac properties. Life at 35,000 feet was simply better in every way.

Few things suggested the sophistication of international travel more potently than a Rolex GMT-Master strapped to a man's wrist. Airline pilot was a high-status profession, and rather like mayoral chains of office, this watch was a vessel in which that status was conveyed. The 2002 film *Catch Me if You Can*, in which Leonardo di Caprio portrays real-life confidence trickster Frank Abagnale, who masqueraded as a Pan Am pilot, is a paean to the lost glamour of commercial air travel. It vividly evokes the excitement and sense of promise that attended jet travel.

In a society that was still largely defined by executive hierarchy and gender stereotypes, airline pilots enjoyed a position that was at once prestigious, well remunerated and glamorous. Firmly grounded deskbound executives with their three-Martini lunches and Cadillacs may have enjoyed the compensations of the Eisenhower boom and the thriving military industrial complex, but they could be forgiven for wanting to appropriate a little bit of 'airline pilot cool' with a Rolex GMT-Master.

There were few places where the social apartheid that divided the airborne from the deskbound was more keenly felt than at Pan Am's headquarters, where watches intended for flight crews were routinely purloined by management. That was until one day, Juan Trippe, the airline's founder, caught sight of a GMT on an executive's wrist and asked why it was not being worn by a pilot. Trippe ordered that all the GMT-Masters should be returned, to be issued solely to flight crews. But, as a concession, he commissioned 100 Rolex GMTs with *white* dials for the so-called 'desk pilots'.

If the Boeing 707 had been responsible for the romance of jet travel, it was another Boeing that devalued the currency of the jet. With the launch of the Boeing 747, the cargo plane for

The Pan-Am building, designed by Walter Gropius, towered over central Manhattan and testified to the importance of the emblematic airline of which Rolex was the official timepiece.

Above: The 2002 film Catch Me If You Can *was a nostalgic paean to the lost era of airborne glamour associated with the Rolex GMT-Master.*

Right: A 1960s magazine ad.

people that shaped the era of mass air travel, the lustre of 'the jet set' began to tarnish; what had once been a heaven of exclusivity, glamour and excitement began its slow descent into the now familiar hell of crowded airports and cramped budget flights.

Today, all that remains true to the memory of those intoxicating days when air travel was the acme of glamour is the Rolex GMT-Master.

1. Catalogue Rolex Oyster Wristwatches, UK Market, circa 1958, Rolex
2. Advertisement, Rolex
3. Trans World Airlines

The Moon's Official Wristwatch

Omega Speedmaster

The weather was unusually cold in Washington on 20 January 1961, but that did not deter the crowds who gathered for the inauguration of the youngest man ever to occupy the Oval Office. Compared to the outgoing president, the Second World War general Eisenhower, and the homely First Lady Mamie, 44-year-old John F. Kennedy and his glamorous wife Jacqueline represented a sort of national rebirth, the dawn of a new USA.

Exactly three months later, the thirty-fifth president of the USA could have been forgiven for believing that that dawn was looking more and more like a sunset. On 20 April he scribbled a hasty signature at the bottom of a memo to his vice-president, the shrewd southerner Lyndon B. Johnson, in his capacity as chairman of the Space Council.

Buzz Aldrin planted the US flag on the moon's surface, like the Omega Speedmaster on his wrist, the flag was also sourced by the indefatigable Jim Ragan.

Buzz Aldrin aboard the Eagle Lunar Module during Apollo 11 mission.

Almost sixty years later, his sense of urgency, desperation even, is easily felt. His first point said it all:

> *Do we have a chance of beating the Soviets by putting a laboratory in space, or by a trip around the moon, or by a rocket to land on the moon, or by a rocket to go to the moon and back with a man. Is there any other space program which promises dramatic results in which we could win?*[1]

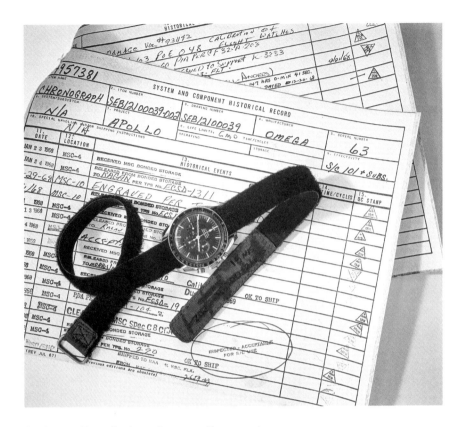

Speedmaster with extra length strap for wear outside a space suit.

On the morning of 14 April, he had woken to the news that the USSR had sent
a man into space and brought him back. In the climate of fear and suspicion that
characterized the years of the Cold War, the world reacted with astonishment that
a supposedly backward country had pulled so far ahead of the USA in what was
being described as 'the space race'. Armchair Cold Warriors reasoned that if the
'Commies' could send a man into space, they would surely find firing an atomic
missile into the middle of the USA simple enough.

Kennedy, who had hitherto regarded space as little more than a distraction, was
shaken. Quivering with the sort of frustration that would have been familiar to King
Henry – 'Will no one rid me of this turbulent priest?' – II of England, he pleaded:
'If somebody can, just tell me how to catch up.' He clutched at the flimsiest hope:
'Let's find somebody – anybody. I don't care if it's the janitor over there.'[2]

It did not help his mood that, as he paced his office that evening demanding answers
from his scientific advisers, a fleet was sailing towards Cuba and the humiliating
fiasco of an invasion attempt that has become known as the Bay of Pigs. By the
afternoon of 19 April, the counter-revolutionary force had been overcome.

First a farmer's boy from Smolensk had beaten the mighty USA into space, and then a bunch of supposedly ill-disciplined, unshaven revolutionaries had pushed the US-backed invasion of Cuba into the sea. Unable to conquer an island just a few dozen miles off the US coast, and with numbers of American troops committed to Vietnam rising steadily, JFK was in search of battle he could win, and he decided that the battle would be fought not in the Caribbean, nor in Indochina – but on the moon.

Today, at a distance of over half a century, the moon landings are historic events, but it is worth considering them in the context of the times. In 1961, when Kennedy decided almost overnight to send men to the moon, it was only six decades since Dumont had piloted his dirigible around the Eiffel Tower, and even when Neil Armstrong took his first steps on the moon, there were still many people alive who could remember a time before Dumont and the Wright brothers had given man wings (incidentally, there is a crater on the moon named in Dumont's honour).

On 25 May, Kennedy made the famous 'Urgent National Needs' speech to a Joint Session of Congress. 'I believe that this nation should commit itself to achieving the goal, before this decade is out, of landing a man on the moon and returning him safely to the Earth' ranks alongside 'Ich bin ein Berliner' among JFK's memorable oratorical flourishes.

Less frequently quoted is the following sentence: 'No single space project in this period will be more impressive to mankind, or more important for the long-range exploration of space; and none will be so difficult or expensive to accomplish.' He went on to describe it as a goal that would touch every person in the USA: 'in a very real sense, it will not be one man going to the moon – if we make this judgment affirmatively, it will be an entire nation. For all of us must work to put him there.'

By 1965, a quarter of a million people were employed on the most ambitious technical project ever undertaken by mankind. Among them was Jim Ragan, who, like LBJ, was as Texan as a steak dinner at a rodeo. Today, Ragan is a raven-haired septuagenarian with commendably cheerful taste in ties and shirts, who favours a hand-tooled leather holster for his mobile phone. His conversation is seasoned with exclamations like 'Jeez!' and his pronunciation of the word 'vehicle' makes the most of what for many is a largely silent 'h'.

In 1964, Ragan was a lanky physics graduate. He tells me: 'I happened to know somebody that got me an interview at NASA,' where he recalls being sent 'around three or four different areas'[3] before ending up across a desk from forty-year-old Deke Slayton, coordinator of astronaut activities – a buzz-cut, craggy-featured Second World War pilot who had been an astronaut in the Mercury programme until being grounded for an irregular heartbeat.

Apollo 13's Jack Swigert wearing an Omega speedmaster around the sleeve of his spacesuit.

Slayton's brief was the definition of wide. He was responsible for everything from selection of astronauts down to the most mundane pieces of equipment and, on the day that he interviewed Ragan, cameras were uppermost in his mind. The pictures from space had hitherto been taken with commercial cameras from a cramped space vehicle. Unsurprisingly, they were terrible. Knowing that his interviewee was a physics graduate, and hoping he knew something about optics, he tossed him a question.

'Do you think you can develop us a camera that we can get good pictures from?'

'Sure I can, why not?' came the answer.

'And so, he basically hired me on that premise,' Ragan says, 'that I was going to do all the cameras. Well, it turned out I ended up doing just about all of the hardware that we provided for the crew. We did a lot of weird stuff like barf bags in case they got sick up there, wet wipes, and that type of thing. Pens, pencils, markers, the whole bit.' [4]

But his first assignment from Slayton was neither a sick bag nor a camera. 'When I got there, he says: "We have these watches, three different ones and they didn't perform very well up there."' [5]

Thus, Ragan set about finding watches. But first he had to devise the tests to simulate the rigours of lunar use… not that anyone in the mid-1960s knew what the rigours of lunar use actually were.

Testing of astronauts and their equipment was brutal and at times remarkably ad hoc: weightlessness training for early cosmonauts in the USSR including being dropped down the lift shaft of Moscow State University in a cage that crashed into buffers of compressed air. No one knew exactly what awaited man when he stepped on the moon, so Ragan devised tests for everything he could think of.

> *The watches were subjected to temperatures ranging from 71°C to 93°C (159–199°F) over a two-day period, after which they were frozen at −18°C (0°F). They were placed in a vacuum chamber heated to 93°C, and then subjected to a test where they were heated to 70°C and then were immediately cooled to −18°C… not once, but fifteen times in rapid succession! When this had been completed, the watches were subjected to 40 G in six different directions and to high and low pressures. Their performance and ability to function was tested in an atmosphere of 93% humidity and in a highly corrosive 100% oxygen environment. The watches even had to endure noise levels reaching 130 decibels. Finally, they were vibrated with average accelerations of 8.8 G.* [6]

Ragan prepared a tender and, out of ten watch companies approached, four submitted watches, one of which did not even make it to the first test on account of its size. 'I could have mounted it on a ship somewhere. It was a monstrous thing, so it automatically got ruled out because it wasn't wrist-worn. The other two got killed in the thermal vacuum test where we ran the temperatures up and down.' [7] The watches that failed featured bi-metallic hands made from materials that did not expand or contract at uniform rates when subjected to extreme temperatures; the result was that the hands distorted and curled around each other.

Below, right and opposite: The Omega Speedmaster had to go through a qualification program before it could be declared 'flight-qualified by NASA for all manned space missions'. This involved testing its capabilities under numerous and challenging atmospheric conditions.

SHOCK MACHINE
CAPACITY
150 POUNDS TO 100 Gs

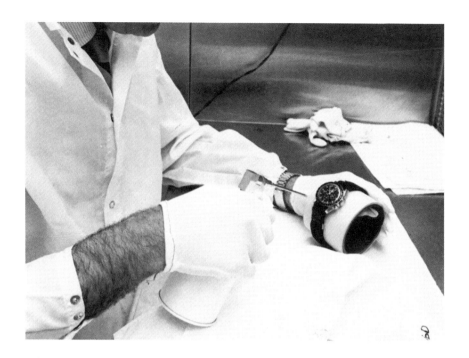

Rather fittingly, the last watch ticking came from a company called Omega, based in the Swiss town of Biel. Its model, the Speedmaster, a chronograph, had been launched in 1957 and had caused a modest sensation in the watch industry because of its calibrated outer bezel, which carried a tachymetric scale that enabled the wearer to calculate speeds and rates of production. 'We designed the Speedmaster for men who reckon time in seconds,' boomed the advertising, 'for scientists, engineers, T.V. and movie directors.' [8]

By March 1965, it could have added 'astronauts' to that list, as the watch was designated 'Flight Qualified for all Manned Space Missions'.

Within weeks it was in space on the wrists of Gemini III astronauts Virgil 'Gus' Grissom and John Young. But the first Omega knew about this extra-terrestrial use of their products was in June 1965, when Edward H. White performed America's first spacewalk as part of the Gemini IV mission with his Speedmaster visible in the photographs (which were, incidentally, much improved since Ragan's arrival). Until then, Omega's US distributor, Norman Morris, which had supplied the watches, was unaware of their intended use: 'that's the first time that Omega knew we were flying their watch.' [9]

It was the ordinariness of the watch that made it extraordinary: few men would walk in space – fewer still would leave their footprints on the moon – yet men around the world could buy and wear exactly the same model of watch that Ed White wore on his EVA, and that Armstrong and Aldrin would wear on Velcro straps outside their space suits when they made their historic lunar landing. 'Anyone could have bought it for $85 a piece, just like I did,' recalls Ragan with a chuckle. [10]

By the mid-1960s, 'NASA's spending would command 5 per cent of the nation's entire annual federal budget', [11] and yet the few men on whom many billions were spent wore watches costing less than $100. 'We bought a total of ninety-seven. That included everything I bought from the beginning of Gemini through the Apollo programme and the Skylab programme,' Ragan recalls. Once the watches had been 'flown', they were returned to him. 'I got them in, had them re-lubed and all on the inside, and if the crystal was scratched, replaced that. The ones that really got messy were the ones that went to the moon. They would come back with this black-looking, graphite-like material on them. It was lunar soil. And if the crystal blew off [there are the remains of three acrylic watch glasses on the moon], it was right down in the movement – but it was still running.' [12]

Omega is understandably proud to boast that 'no other piece of equipment – let alone a watch – can claim to have been used during the Mercury, Gemini, Apollo, Skylab, Soyuz, Salyut, Space Shuttle, MIR and International Space Station programs'.

The Speedmaster contributed to one lunar mission in particular: the successful failure known as Apollo 13. The mission was aborted after an explosion on the outward journey damaged the spacecraft and forced the crew to seek refuge in the Lunar Module. After going once round the moon, they headed back to Earth with the command module, using the Speedmaster to time the crucial engine firings necessary for their safe return.

Space has made the Omega Speedmaster famous – and valuable: in May 2018, an Omega Speedmaster fetched the equivalent of $400,000 at auction. It also helped make history and, in the case of Apollo 13, helped Kennedy to make good on his promise not just to put a man on the moon – but to bring him home as well.

When Kennedy said, 'all of us must work to put him there', it turned out that he was not just talking of the USA, but the Omega factory in Biel as well.

1. 'Memo from President John F. Kennedy to Vice President Lyndon Johnson, April 20, 1961', National Archives and Records Administration, Lyndon Baines Johnson Library and Museum, Austin, Texas (https://www.visitthecapitol.gov/exhibitions/artifact/memo-president-john-f-kennedy-vice-president-lyndon-johnson-april-20-1961)
2. Quoted from Jamie Dornan and Piers Bizony, *Starman: The Truth Behind the Legend of Yuri Gagarin* (London: Bloomsbury, 1998), p. 142
3. Interview with the author, December 2018
4. *Ibid.*
5. *Ibid.*
6. The OMEGA Speedmaster and the World of Space Exploration, a pamphlet published by Omega, pp. 6–7
7. Interview with the author, December 2018
8. Grégoire Rossier and Anthony Marquié, *Moonwatch Only: 60 Years of OMEGA Speedmaster* (Watchprint, 2014), p. 239
9. *Ibid.*
10. *Ibid.*
11. Jamie Dornan and Piers Bizony, *Starman, op. cit.* 144
12. Interview with the author, December 2018

Glossary

Automatic a watch that uses the circular motion of a weight caused by the wearer's arm movements to rewind the mainspring, also known as self-winding

Balance spring spring attached to the balance wheel that controls the oscillations

Bezel part of a watch case that surrounds the glass: sometimes calibrated (Omega Speedmaster), sometimes moveable (Rolex GMT-Master)

Breguet curve curvature at the end of the balance spring to improve performance, made famous by its eponym Abraham Louis Breguet

Carillon system of bells that when struck in order plays a tune

Chronograph stopwatch-type function incorporated into a pocket or wristwatch

Chronometer a timepiece of great accuracy. The movement is tested by an independent authority such as the COSC (Controle Officiel Suisse de Chronometres) to perform within the parameters of a small number of seconds over several days in several positions and at varying temperatures

Complication any horological function beyond the display of the time in hours, minutes and seconds, for example an indicator showing the phases of the moon or a date window

Deadbeat escapement a highly accurate type of escapement perfected by George Graham, mentor of John Harrison

Equation of time difference between mean time (i.e. that shown on a watch) and solar time (i.e. that shown on a sundial). Mean and solar time coincide four times a year: because of the Earth's elliptical orbit around the sun, there is a fluctuating difference between mean and solar time for the rest of the year

Equinox Twice-yearly occasion on which day and night are of equal length (*see* solstice)

Escapement Part of a watch or clock that turns the continuous energy released by the mainspring of a watch or descending weights of a clock into little impulses marking units of time; it is the escapement that creates the characteristic ticking noise made by mechanical timepieces

Foliot a bar or arm at each end of which is a weight that can be moved to speed up or slow down the time, the foliot is mounted on a slender shaft called a verge with two protrusions which catch the teeth of the wheel moved by the falling weights on rope. This creates the familiar ticking sound, while the position of the weights regulates the speed at which the verge turns, transmitting energy to the indication of time. The verge and foliot system, a key component of early mechanical clocks, was later superseded by the pendulum. At first the pendulums swung through a wide arc of up to 100 degrees, but the invention of the 'anchor' escapement limited the swing to as little as four or five degrees, making the clock more accurate and enabling long pendulums to fit within longcase (grandfather) clocks

Fusee a grooved cone around which is wound a chain also connected to the barrel with the mainspring. As the spring unwinds, it wraps itself with the chain from the fusee, which, thanks to its conical shape, offers different levels of resistance evening out the delivery of energy to the movement. A key component of early watches, the principle is similar to that of derailleur gears on a bicycle

Gear train system of interconnected toothed components (wheels and pinions) which turn upon each other, and through which power and motions are transmitted; for instance when a mechanical watch is wound, a gear train transmits the energy from the winding crown moved by thumb and index finger to the mainspring, where it is stored and released using another gear train called the drive train

Gravity escapement the double three-legged gravity escapement (also known as the Grimthorpe escapement) is a constant force device that has the goal of delivering equal amounts of power to the escapement. Its design also isolates the functions of escapement as far as possible – from the external factors that might affect a tower clock, such as the effect of high winds, and the weight of ice and snow on hands, which can be fed back through the movement, altering the performance. The double three-legged gravity escapement also functions without lubrication, another benefit in the environment of a tower clock, where extremes of temperature affect oil viscosity

Jacquemart mechanical figure driven by clockwork to strike the bell on a clock

Latitude a series of notional lines running parallel to the equator: combined with longitude to provide a map reference

Longitude a series of notional lines running perpendicularly from pole to pole that slice the world into segments reminiscent of an orange: combined with latitude to provide a map reference

Mainspring the spring in a watch or clock that, once wound, unwinds gradually, powering the movement. As an alternative to weight-driven clocks, the mainspring enabled the development of the watch, which for obvious reasons could not use weights

Meridian a notional line that passes from north to south poles via a specific place, for instance Greenwich, London

Perpetual calendar a function that enables a watch to give the day, date and month without needing to be reset to take account of leap years

Repeater a watch that strikes the time when a slide or pusher is activated. A minute repeater strikes the hours with a low tone, the quarters with a mixture of high and low tones and minutes with a rapid succession of high notes

Sidereal time sidereal time is based on the rotations of the Earth relative to stars so distant that they appear not to move. As well as performing one rotation about its own axis in a day, the Earth also performs approximately a single degree of its annual orbit around the sun, meaning that solar noon will be at a slightly different time from day to day; sidereal days are about four minutes shorter than solar days

Solstice twice-yearly occasion when the difference between the length of day and night is greatest

Tourbillon literally translated as whirlwind, the tourbillon was invented by Abraham Louis Breguet. The escapement is placed in a rotating cage to compensate for the effects of gravity when the watch is in a vertical position (for example in the pocket of its wearer), and this mechanical refinement became a horological fashion during the late twentieth and early twenty-first century for wristwatches

Select Bibliography

BOOKS

Al-Jazari, Ibn. *The Book of Knowledge of Ingenious Mechanical Devices*. Boston: Dordrecht-Holland, 1974

Arbor, Ann. *Time: Histories and Ethnologies*. University of Michigan Press, 1995

Bardenhewer, Otto. *Patrology: The Lives and Works of the Fathers of the Church 1851–1935*

Bartky, Ian R. *The Adoption of Standard Time, Technology and Culture*. London: The Johns Hopkins University Press and the Society for the History of Technology, 1989

Berner, G.-A. *Dictionnaire professionnel illustré de l'horlogerie*, Vols. 1 and 2. Bienne: Société du Journal La Suisse Horlogère SA, 1961

Breguet, Emmanuel. *Breguet Watchmakers, Since 1775*. Paris: Gourcuff, 1997

Breguet, Emmanuel. *Art and Innovation in Watchmaking*. Prestel, 2015

Brouria Bitton-Ashkelony & Aryeh Kofsky Koninklijke Brill NV. *Christian Gaza in Late Antiquity*. Boston: Brill Leiden, 2004

Chapuis, Alfred and Gélis, Edouard. *Le Monde des automates: étude historique et technique*, Vol. 2. Geneva: Slatkine, 1928

Collins, Wilkie. *Armadale*. London: Penguin Classics, 1866

Cologni, Chaille, Flechon. *The Beauty of Time*. Paris: Flammarion, 2018

Conihout, Isabelle de and Fritsch, Julia., *Ces Curieux Navires: Trois Automates de La Renaissance*. Paris: Réunion des Musées Nationaux, 1999

Davies, Norman. *Europe: a History*. Bodley Head, 2014

Dohrn-van Rossum, Gerhard. *History of the Hour: Clocks and Modern Temporal Orders*. University of Chicago Press, 1996

Flechon, Dominique. *The Mastery of Time*. Paris: Flammarion, 2011

Fleming, Sir Sandford. *Time-reckoning for the Twentieth Century*. Montreal: Dawson Bros Montreal, 1886

Fleming, Sir Sandford. *Terrestrial Time, A memoir*. 1876

Forsyth, Hazel. *The Cheapside Hoard: London's Lost Jewels*. London: Philip Wilson Publishers, 2013

Foulkes, Nicholas. *The Impossible Collection of Watches*. Assouline, 2014

Foulkes, Nicholas. *Patek Philippe: The Authorized Biography*. London: Preface, 2016

Foulkes, Nicholas. *Automata*. Editions Xavier Barral, 2017

Fritz, Manfred. *Reverso – The Living Legend*. Jaeger-LeCoultre, Edition Braus,1992

Gibbon, Edward. *The History of the Decline and Fall of the Roman Empire*. London: Strahan & Cadell, 1808

Goodall, John. *A Journey in Time the Remarkable story of Seiko*, Good Impressions. United Kingdom, Hertfordshire, 2003

Hill, Rosemary. *God's Architect: Pugin and the Building of Romantic Britain*. Yale University Press, 2007

Howse, Derek. Warrant. *Greenwich Time and the Discovery of the Longitude*. Oxford University Press, 1980

Lucas, J. R. *A Treatise on Time and Space*. 1973, Part 1

MacDonald, Peter. *Big Ben: The Bell, The Clock and The Tower*. Gloucester: History Press, 2005

Marchant, Jo. *Decoding the Heavens, Solving the Mystery of the World's First Computer*. Windmill Books, 2009

Marshack, Alexander. *Cognitive Aspects of Upper Paleolithic Engraving*. University of Chicago Press on behalf of Wenner-Gren Foundation for Anthropological Research, 1972

Marshall, Peter. *The Mercurial Emperor: The Magic Circle of Rudolf II in Renaissance Prague*. London: Vintage Books, 2013

Morton, H. V., *In Search of England*. London: Methuen & Co. Ltd., 2000

Morus, Iwan Rhys. *The Oxford Illustrated History of Science*. Oxford University Press, 2017

Needham, Joseph and Ling, Wang. *Science and Civilisation in China, Volume 4: Physics and Physical Technology Part II: Mechanical Engineering*. Cambridge University Press, 1971

North, John. *God's Clockmaker: Richard of Wallingford and the Invention of Time*. London: Bloomsbury, 2005

Patek Philippe Watches, Vol. 1 and 2. Geneva: Patek Philippe Museum, 2013

Pelletier, Alain. *Boeing, The Complete Story*. London: Haynes Publishing, 2010

Perman, Stacy. *A Grand Complication: The Race to Build the World's Most Legendary Watch*. London: Atria Books, 2002

Robinson-Planche, James. *A History of British Costume*. London: Charles Knight, 1834

Rossier, Grégoire and Marquié, Anthony., *Moonwatch Only: 60 Years of OMEGA Speedmaster*. Watchprint, 2014

Salomons, Sir David Lionel. *Breguet*. London, 1921

Santos-Dumont, Alberto. *My Airships: the story of my life*. London: Grant Richards: The Riverside Press Limited, 1904

Scholz, B. Walter *Carolingian chronicles: Royal Frankish Annals and Nithard's Histories*. University of Michigan Press, 1970

See Ye, Shilin Yanyu. *The Stone Forest*. Beijing: Zhonghua Shuju, 1984

Sheppard, Francis. *The Treasury of London's Past: An Historical Account of the Museum of London and Its Predecessors, the Guildhall Museum and the London Museum*. London: HMSO, 1991

Sobel, Dava and Andrewes, William J.H. *The illustrated Longitude*. New York City: Walker & Co., 2003

Turnbull, Stephen. *The Samurai and the Sacred: The Path of the Warrior*. London: Osprey, 2009

Vincent, Clare and Leopold, Jan Hendrik. *European Clocks and Watches In the Metropolitan Museum of Art*. Yale University Press, 2015

Weinert, F. *The March of Time; Evolving Conceptions of Time in the Light of Scientific Discoveries.* Springer, 2013

White, Edmund. *Arts and Letters.* New Jersey: Cleis Press, 2006

Withers, Charles W.J. *Zero Degrees: Geographies of the Prime Meridian.* Harvard University Press, 2017

JOURNAL ARTICLES

Bedini, Silvio. A. and Maddion, Francis. R. *Mechanical Universe: The Astrarium of Giovanni de' Dondi.* Transactions of the American Philosophical Society, Vol. 56, No. 5 (1966)

Critical Inquiry (1977), p101

Hendry, Joy "Time in a Japanese Context". Exhibition Catalogue 'The Story of Time (1999)

De Maisieres, Tome XVI, pp. 227–22

De Solla Price, Derek. "Gears from the Greeks. The Antikythera Mechanism: A Calendar Computer from ca. 80" *Transactions of the American Philosophical Society* Vol. 64, No. 7 (1974), pp. 1–70

Haber, F.C. "The Cathedral Clock and the Cosmological Clock Metaphor in The Study of Time II", pp. 399–416

Hunt, J L. "The Handlers of Time: The Belville Family and the Royal Observatory, 1811–1939". *Astronomy & Geophysics*, Vol. 40, Issue 1 (1999)

Kelly, George Armstrong. "The Machine of the Duc D'Orléans and the New Politics." *The Journal of Modern History*, Vol. 51, No. 4 (1979) pp. 667–84

Liu, Heping. "Northern Song Imperial Patronage of Art, Commerce, and Science". *The Art Bulletin*, Vol. 84, No. 4 (2002), pp. 566–95

Marquet, Louis. "Le Canon Solaire du Palais-Royal à Paris". *L'Astronomie*, Vol. 93, 1979, p. 369

Popular Scientist. October 1929, p. 63

Poulle, Emmanuel. "L'horlogerie a-t-elle tué les heures inégales, Bibliothèque de l'École des chartes

Vol. 157, No. 1, "Construire Le Temps: Normes et Usages Chronologiques au Moyen Âge (janvier-juin 1999)", pp. 137–56. Published by Librairie Droz

Rolex Archives. *Catalogue Rolex Oyster Wristwatches UK Market.* (1958)

Smith, Roff. *National Geographic.* (2013)

Soppelsa, Peter and Stern, Blair. "Santos-Dumont's Blimp Passes the Eiffel Tower. Source: Technology and Culture", Vol. 54, No. 4 (October 2013), pp. 942–46, published by: The Johns Hopkins University Press and the Society for the History of Technology

Sotheby's, Important Watches, Including the highly important Henry Graves JR Supercomplication. Geneva, 11 November 2014

Uchida, Hoshimi. "The Spread of Timepieces in the Meiji Period." *Japan Review*, No. 14 (2002). pp. 173–92

Patek Philippe, Geneve. Star Calibre 2000. Editions Scriptar SA

OMEGA, Speedmaster, Press information, 2015

Cartier in Motion, Ivory Press, 2017 Curated by Norman Foster. Authors: Jean-Pierre Blay, Alain de Botton, Bob Colacello, Norman Foster, Nicholas Foulkes, Carole Kasapi, Rossy de Palma, Pierre Rainero and Deyan Sudjic

NEWSPAPER AND MAGAZINE ARTICLES

"Romance and the Colour of London" in *The Times*. London. March 19, 1914

Bitsakis, Yanis. "On Time". *Vanity Fair*, autumn 2012

Girouard, Mark. "Rout to Revolution: The Palais-Royal, Paris". *Country Life*, January 1986

Illustrated London News, March 6 1858

New York Times, November 20, 1983

The New Yorker, April 22, 1974

The Times, London. January 22, 2005

The Times, London. December 13, 1943

Vanity Fair "On Time" (all editions)

WEBSITES

Bank of England Inflation Calculator. https://www.bankofengland.co.uk/monetary-policy/inflation/inflation-calculator

Britannica. www.Britannica.com

Ferguson, James. Account of Franklin's Three-Wheel Clock, [1758], *Founders Online*, National Archives, last modified 13 June, 2018, http://founders.archives.gov/documents/Franklin/01-08-02-0060 [Original source: *The Papers of Benjamin Franklin*, vol. 8, *April 1, 1758, through December 31, 1759*, ed. Leonard W. Labaree. New Haven and London: Yale University Press, 1965, pp. 216–220.]

Royal Museums Greenwich. "Rehabilitating Nevil Maskelyne – Part Four: The Harrisons' accusations, and conclusions". https://www.rmg.co.uk/discover/behind-the-scenes/blog/rehabilitating-nevil-maskelyne-part-four-harrisons-accusations-and

Science Museum. Rooney, David. "Ruth Belville: The Greenwich Time Lady" https://blog.sciencemuseum.org.uk/ruth-belville-the-greenwich-time-lady/. October 2015

Seiko Museum https://museum.seiko.co.jp/en/knowledge/wadokei/variety/

Smithsonian www.smithsonian.com

University of Bologna https://www.unibo.it/en/university/who-we-are/our-history/university-from-12th-to-20th-century

World Monuments Fund https://www.wmf.org/project/mortuary-temple-amenhotep-iii

ARCHIVES

Breguet, Cartier, Omega, Patek Philippe, Rolex

INTERVIEWS

BBC News 15 July 2013, Vince Gaffney interviewed by Huw Edwards

Interview with Wolfram Koeppe, Marina Kellen French Curator, European Sculpture and Decorative Arts at The Metropolitan Museum of Art. June 2018

Interview with Daryn Schnipper, Sothebys: Senior Vice President, Chairman, International Watch Division, New York. January 2019

January 2019: Interview with Pierre Rainero, Cartier, January 2019

Index

C

Critz, John de the Elder *116*
Ctesibius of Alexandria 33, *34*
Current Anthropology 19

D
Däniken, Erich von 19
Dark Ages 57, 83, 92
Das unsterbliche Herz (film) *103*
Dasypodius, Conrad 86, 87, 88, 91
Davies, Norman 55
de la Mare, Thomas 79
deadbeat escapement 234
Dee, John 108
Dee, River 22, *24*, 25
DeLille, Jacques 142
Denison, Edmund Beckett 166–7, 168, 171, 173
Dent, Edward John 166, 167, 168, 170, 171, 173
Dent, Frederick 170, 171
Dental News 194
Desoutter, Louis 154
Deutsch Prize 184–6, *185*, 187, 189, 190, *192*
di Caprio, Leonardo 213, *215*
Dickins, Charles 90
Diels, Hermann *51*
Diller, Na'aman 155
Dion, Count de 186
Dodecanese islands 36–8, *41*
Dodge brothers 203
Don, River 22, *24*
Dondi, Giovanni de' (Master John) 92–9, *92*, *93*, *98*
Dondi, Jacopo de' 83, 92, 96–7
Dowd, Charles Ferdinand 178–9, *178*, 183
Dubai 68

E
Earnshaw, Thomas 136
East India Company 117, 134
Edo period (1603–1868) 122–4, *123*, 126
Edward, Lake 15, *17*, *18–19*
Edward I 76
Edward II 77
Edward III 77, 79
Egypt 7, 8
    Karnak Clepsydra *8*, 28–33, 59
Eiffel Tower 11, 184, *184*, 187, 219
Eisenhower, President 214, 216
Elephant Clock of Al-Jazari 68–73, *69*
Elizabeth I, Queen 108, 117, 191
Elizabeth II, Queen 173
emerald watch of Cheapside Hoard 112–19, *112*

Ephebe *43*
equation of time 234
equinox 49, 91, 234
escapements 234
    deadbeat escapement 234
    double three-legged gravity escapement 168–9, *168*, 235

F
factory clocks 9
farming calendars *47*
Ferguson, James 146–7
Ferguson's Clock 144–6
Ferro Meridian 179
Fersen, Axel von 152, 154, *154*
Flamsteed, John 133
Fléchon, Dominique 54
Fleming, Sandford 174–81, *175*, 183
foliot 75, 124, 234
    double foliot balance 124, *124*
Forsyth, Hazel 116, 117, 118
Fragonard, Jean-Honoré 140
Francesco I of Milan *95*, 98
Franklin, Benjamin 144–7, *147*
Franklin, John 144
Franks 57, 59
French Revolution 142–3, 150, 157
Frick, Henry Clay 203
Fritsch, Julia 111
Frodsham 203
fusee 234–5
Futurism 190–1

G
Gaffney, Professor Vincent 22, 23, 24, *24*, 25, 27
Galileo 149
gamma-radiography 41–2
Garros, Roland 192
Gaunt, Ben 169
Gaza, Great Clock of 50–5, 57, 82, 86
gear train 235
Gemini space missions 223, 224, 225
Genoa, Doge of 97
George II, King 135
George III, King 135, 136, 161
George V, King 115
Gesner, Andreas, *Julius Caesar 44*
Gibbon, Edward 56, *57*
Gishodo Suwako Watch and Clock Museum *62*
GMT (Greenwich Mean Time) 181
GMT-Master 210–15

Picture Credits

Jacket: **FC** = front cover, **BC** = back cover, **t** = top, **b** = bottom, **c** = centre, **a** = above, **l** = left, **r** = right: **FC cr, Spine ca** Alamy / Rade Lukovic; **FC cla** Arcangel Images / Sybille Sterk; **FC ca, Spine bl** Bridgeman Images / © Florilegius; **BC l** Bridgeman Images / Private Collection; **FC t, BC c** Bridgeman Images / The Worshipful Company of Clockmakers' Collection; **FC bc, Spine t** Shutterstock / donatas1205; **FC and BC framework** Shutterstock / OS; **BC r** Shutterstock / pandapaw.

Pages: **Pg 1** © Collection Montres Breguet SA; **Pg 2** 123RF / Sebastien Coell; **Pg 3** Courtesy Sotheby's; **Pg 7** Getty Images / De Agostini Picture Library; **Pg 8** Getty Images / Science & Society Picture Library; **Pg 10** Alamy / Peter Schickert; **Pg 12-13** Getty Images / AFP/ Fabrice Coffrini; **Pg 14** Alamy / GFC Collection; **Pg 16** National Geographic / Alexander Marshack; **Pg 17t** The Royal Belgian Institute of Natural Sciences, Brussels; **Pg 17b** Courtesy of the Peabody Museum of Archaeology and Ethnology, Harvard University, PM2005.16.2.318.38; Gift of Elaine F. Marshack, 2005; **Pg 18-19** Hicham Baoudi; **Pg 20** The Royal Belgian Institute of Natural Sciences, Brussels; **Pg 22** © Murray Archaeological Services; **Pg 23** © Crown Copyright: HES, Image No. SC 1071719; **Pg 24** University of Birmingham; **Pg 25** Illustration courtesy Eugene Ch'ng, Eleanor Ramsey and Vincent Gaffney; **Pg 26** © Murray Archaeological Services; **Pg 27** V. Gaffney et al. 2013 *Time and a Place: A luni-solar "time-reckoner" from 8th millennium BC Scotland*, Internet Archaeology 34. https://doi.org/10.11141/ia.34.1; **Pg 28** Shutterstock / Zbigniew Guzowski; **Pg 29** Alamy / Heritage Image Partnership Ltd; **Pg 30** Getty Images / Werner Forman / UIG; **Pg 31** Getty Images / Science & Society Picture Library; **Pg 32** Alamy / Photo 12; **Pg 33** TopFoto.co.uk; **Pg 34** Alamy / The Print Collector; **Pg 35** Alamy / Chronicle; **Pg 36** Antikythera Mechanism Research Project; **Pg 37** Alamy / Nature Picture Library; **Pg 38** Getty Images / AFP/ Louisa Gouliamaki; **Pg 39** Alamy / Have Camera Will Travel | Europe; **Pg 40** Alamy / The History Collection; **Pg 41** Alamy / Abbus Acastra; **Pg 42** Photo by Malcolm Kirk, courtesy of the de Solla Price family; **Pg 43** Getty Images / Leemage; **Pg 44** Getty Images / Heritage Images / Historica Graphica Collection; **Pg 45** Getty Images / DeAgostini; **Pg 46** Alamy / Vito Arcomano; **Pg 47** AKG / Rabatti & Domingie; **Pg 48** Getty Images / De Agostini / DEA / G. Dagli Orti; **Pg 51** Hermann Diels, *Über die von Prokop beschriebene Kunstuhr von Gaza. Mit einem Anhang enthaltend Text und Übersetzung der ἐκφρασις ὡρολογιον des Prokopios von Gaza*, 1917; **Pg 52-53** Alamy / Christine Webb; **Pg 56** Wikimedia Commons https://commons.wikimedia.org/wiki/File:Wasseruhr_Harun_al_Raschid.jpg; **Pg 57** Alamy / Heritage Image Partnership Ltd; **Pg 58** Getty Images / Royal Geographical Society; **Pg 60** Joseph Needham, Science and Civilization in China: Volume 4, Part 2, Mechanical Engineering, page 451; **Pg 61** Getty Images / Science & Society Picture Library; **Pg 62** © Maya Vision International Ltd; **Pg 63** Getty Images / Science & Society Picture Library; **Pg 64** akg-images / Pictures From History; **Pg 66** Bridgeman Images / Pictures from History; **Pg 69** Metropolitan Museum of Art / Bequest of Cora Timken Burnett, 1956 / Acc No 57.51.23; **Pg 70** Metropolitan Museum of Art / Rogers Fund, 1955 / Acc No 55.121.11; **Pg 71** Metropolitan Museum of Art / Rogers Fund, 1955 / Acc No 55.121.12; **Pg 72** Alamy / Tuul and Bruno Morandi; **Pg 74** Bridgeman Images / British Library; **Pg 75** Bridgeman Images / British Library; **Pg 76** Bridgeman Images / Photo © John Bethell; **Pg 78** 123RF / Sebastien Coell; **Pg 80** © Bodleian Libraries, University of Oxford MS. Laud Misc. 657 fol. 047r; **Pg 82** 123RF / meinzahn; **Pg 83** Alamy / INTERFOTO; **Pg 84** Metropolitan Museum of Art / Anonymous Gift, 2009. Acc No 2009.157; **Pg 87** 123RF / Olena Kachmar; **Pg 88** Getty Images / DEA / G. Dagli Orti; **Pg 89** Wilkie Collins, *Armadale*, 1866, page 221; **Pg 90** akg-images / Massimiliano Pezzolini; **Pg 92** Alamy / Science History Images; **Pg 93** Getty Images / Science & Society Picture Library; **Pg 94** Getty Images / Fototeca Storica Nazionale; **Pg 95** Alamy / Heritage Image Partnership Ltd; **Pg 96** Alamy / Peter Horree; **Pg 98** 123RF / sedmak; **Pg 100** www.peterhenlein.de; **Pg 101t** Getty Images / Harold Cunningham; **Pg 101b** Alamy / Panther Media GmbH; **Pg 102t** Stamp of the Third Reich, Germany 1942 MNH commemorating the 400th Anniversary of the Death Peter Henlein; **Pg 102b** Alamy / Gibon Art; **Pg 103** Getty Images / ullstein bild Dtl; **Pg 105** Getty Images / Fine Art Images / Heritage Images; **Pg 106-107** © Trustees of the British Museum; **Pg 109** Photo © RMN-Grand Palais (musée de la Renaissance, château d'Ecouen) / Hervé Lewandowski; **Pg 110** Photo © RMN-Grand Palais (musée de la Renaissance, château d'Ecouen) / Hervé Lewandowski; **Pg 112l** and **r** © Museum of London; **Pg 114-115** Alamy / Chronicle; **Pg 116** Alamy / Chronicle; **Pg 118** Getty Images / Bloomberg; **Pg 119** Thomas Midelton, *A Chast Mayd in Cheapeside*, frontis,1630; **Pg 120** Bridgeman Images / Museu de Marinha, Lisbon, Portugal / Photo © Selva; **Pg 121** akg-images / Fototeca Gilardi; **Pg 122** Alamy / World History Archive; **Pg 123** akg-images / Interfoto / Antiquariat Felix Lorenz; **Pg 124** Seiko Museum; **Pg 125** Library of Congress LC-DIG-jpd-00114; **Pg 126** Seiko Museum; **Pg 127** Alamy / World History Archive; **Pg 128-130** © National Maritime Museum, Greenwich, UK, Ministry of Defence Art Collection; **Pg 131** Getty Images / Science & Society Picture Library; **Pg 133** © National Maritime Museum, Greenwich, London, presented by the descendants of Nevil Maskelyne; **Pg 134** Getty Images / Science & Society Picture Library; **Pg 137** Getty Images / Universal Images Group; **Pg 138** Getty Images / Imagno; **Pg 139** Photo ©RMN-Grand Palais (MuCEM) / Jean-Gilles Berizzi; **Pg 140** Metropolitan Museum of Art / The Elisha Whittelsey Collection, The Elisha Whittelsey Fund, 1961, Acc No: 61.53; **Pg 141** Getty Images / Leemage; **Pg 142** Alamy / Florilegius; **Pg 144-145** Metropolitan Museum of Art. Purchase, Anna-Maria and Stephen Kellen Acquisitions Fund, in honor of Wolfram Koeppe, 2015.; **Pg 146** Wellcome Collection, CC BY (https://wellcomecollection.org/works/p3wk83cw); **Pg 147** Architect of the Capitol, US Capitol; **Pg 148** Getty Images / David Silverman; **Pg 149** Rijksmuseum, Amsterdam; **Pg 150** akg-images / picture-alliance; **Pg 151** © Collection Montres Breguet SA / Xavier Reboud; **Pg 152** © Collection Montres Breguet SA; **Pg 153** Getty Images / Gali Tibbon; **Pg 154** Alamy / Heritage Image Partnership Ltd; **Pg 156** Alamy / Chronicle; **Pg 157t** 123RF / Victoria Demidova; **Pg 157 r** *Popular Science Monthly*, October 1929 p63; **Pg 158** National Maritime Museum, Greenwich, London; **Pg 159** Getty Images / Fox Photos; **Pg 160** Bridgeman Images / The Worshipful Company of Clockmakers' Collection, UK; **Pg 162** Bridgeman Images / Private Collection / Look and Learn / Peter Jackson Collection; **Pg 163** Alamy / Elizabeth Whiting & Associates; **Pg 164** Getty Images / Oli Scarff; **Pg 165 t** Alamy / Chronicle; **Pg 165 b** Alamy / Art Collection 3; **Pg 166** Wikimedia Commons / Morgan & Kidd / https://commons.wikimedia.org/wiki/File:George_Biddell_Airy_1891.jpg; **Pg 167** Alamy / Jon Arnold Images Ltd; **Pg 168** from Vaudrey Mercer, *The Life and Letters of Edward John Dent, chronometer maker, and some account of his successors*. Antiquarian Horological Society,1977; **Pg 169** Getty Images / Daniel Berehulak; **Pg 170** Alamy / The Picture Art Collection; **Pg 171** Alamy / World History Archive; **Pg 172** Alamy / Antiqua Print Gallery; **Pg 175** Science Photo Library / Miriam and Ira D. Wallach Division of Art, Prints and Photographs / New York Public Library; **Pgs 176, 177** Getty Images / Science & Society Picture Library; **Pg 178** Library of Congress LC-USZ62-42699; **Pg 180** Alamy / World History Archive; **Pg 182** Shutterstock / BeautifulBlossoms; **Pg 184** Getty Images / Popperfoto / Bob Thomas; **Pg 185** Getty Images / UniversalImagesGroup; **Pg 186** Vincent Wulveryck, Collection Cartier © Cartier; **Pg 187** Topfoto / Roger-Viollet; **Pg 188** Cartier / © Paul Tissandier; **Pg 191** *Revista Moderna* No 30 April 1899 Anno III p252; **Pg 192** Getty Images / Print Collector; **Pg 193** Archives Cartier Paris © Cartier; **Pg 195** Courtesy Jaeger-LeCoultre; **Pg 196 tr** Getty Images / Bettmann; **Pg 196 l & br** © Antiquorum Genève SA; **Pg 197** Jaeger-LeCoultre Reverso Patent 1931 @Jaeger-LeCoultre Patrimony; **Pgs 199-205** Courtesy Sotheby's; **Pg 206** Alamy / Chronicle; **Pg 208** Getty Images / AFP / Fabrice Coffrini; **Pg 210** Getty Images / Museum of Flight Foundation; **Pg 211-213** Rolex; **Pg 214** Getty Images / F. Roy Kemp; **Pg 215 t** Alamy / Pictorial Press Ltd; **Pg 215 b** Advertising Archive; **Pg 216** NASA; **Pg 217** Alamy / RGB Ventures / SuperStock; **Pgs 218-223** NASA.

Every effort has been made to find and credit the copyright holders of images in this book. We will be pleased to rectify any errors or omissions in future editions.

Acknowledgements

The first and greatest thanks are due to Ian Marshall, who saw hidden in the proposal I sent him the idea for a book, a book that he subsequently did so much to bring into being: without his vision this project would not have come into being. I owe him and his colleagues at Simon and Schuster a great debt and I am truly grateful for their belief and encouragement. Luigi Bonomi my literary agent of many decades is a trusted friend and wise counsellor, a man of saintly disposition who tolerates my idiosyncrasies and does what he can to allay my insecurities. I am grateful for the effort and persistence of my assistant Venetia Stanley, her tirelessness has made her a familiar, if slightly feared, figure in museums, archives and universities around the world.

Among the curators, experts and historians who have been particularly helpful are: Wolfram Koeppe, Marina Kellen, Curator of European Sculpture at The Metropolitan Museum of Art; Daryn Schnipper, Sothebys: Senior Vice President, Chairman, International Watch Division; Professor Vincent Gaffney of the University of Bradford, Hazel Forsyth, Curator at the Museum of London, Patrick Semal, Curator of Anthropology at the Royal Belgium Institute of Natural Sciences, Abdelrahman Othman the curator for the National Museum of Egyptian Civilization in Cairo and his extensive team.

Given that I have spent so much of my life around timepieces I suppose I really ought to thank the watchmaking industry for just being there and continuing to make the beautiful and remarkable objects that are the descendants of the pieces I mention in these pages. I am grateful to Patek Philippe, and Philippe Thierry Stern not only for insight into Patek's own historical pieces but for the exceptional museum devoted to the history of the portable timepiece which anyone passing through Geneva has a duty to visit. On the other side of the world Seiko has assembled a fine collection of timepieces in its museum and I am grateful to its staff for their help with the chapter on Japanese time during the Edo period. I must also thank the archive departments of Rolex, Cartier, Jaeger LeCoultre, Omega and Breguet for their assistance.

Forgive me for not mentioning each individual by name but among the very many people in the industry who have been incredibly kind and helpful during the time I was engaged on this project are, in alphabetical order, Raynald Aeschlimann, Aurel Bacs, Jean Claude Biver, Arnaud Boetsch, Nicolas Bos, Christophe Carrupt, Virginie Chevailler, Laurent Feniou, Kristen Fleener, Jack Forster, Sibylle Gallardo Jammes, Isabelle Gervais, the Hayek family, Annie Holcroft, Wei Koh, Marine Lemonnier, Fabienne Lupo, Sarah Nunziata, Petros Protopapas, Pierre Rainero, John Reardon, Catherine Renier, Karl-Friedrich and Caroline Scheufele, Jasmina Steele, Cyrille Vigneron, Davide Traxler, Patrick Wehrli, Robert Wilson.

First published in Great Britain by Simon & Schuster UK Ltd, 2019
A CBS COMPANY

Copyright © Nicholas Foulkes, 2019

The right of Nicholas Foulkes to be identified as the author of this work has been asserted in accordance with the Copyright, Designs and Patents Act, 1988.

Editorial Director: Ian Marshall
Project Editor: Laura Nickoll
Design: Keith Williams, sprout.uk.com
Picture Researcher: Liz Moore

1 3 5 7 9 10 8 6 4 2

Simon & Schuster UK Ltd
1st Floor
222 Gray's Inn Road
London WC1X 8HB

www.simonandschuster.co.uk
www.simonandschuster.com.au
www.simonandschuster.co.in

Simon & Schuster Australia, Sydney
Simon & Schuster India, New Delhi

The author and publishers have made all reasonable efforts to contact copyright-holders for permission, and apologise for any omissions or errors in the form of credits given. Corrections may be made to future printings.

A CIP catalogue record for this book is available from the British Library

ISBN: 978-1-4711-7064-5
eBook ISBN: 978-1-4711-7065-2

Printed in China

MIX
Paper from
responsible sources
FSC® C104723
FSC
www.fsc.org